Macrophages

Edited by Hridayesh Prakash

Published in London, United Kingdom

IntechOpen

Supporting open minds since 2005

Macrophages
http://dx.doi.org/10.5772/intechopen.87474
Edited by Hridayesh Prakash

Contributors
Dhruv Kumar, Sibi Raj, Vaishali Chandel, Soraya Mezouar, Jean-Louis Mege, Yu-Sheng Wu, Tsung-Meng Wu, Kuang-Teng Wang, Hisang-Lin Tsai, Fan-Hua Nan, Alice Grigore, Hridayesh Prakash, Vishwas Tripathi, Aklank Jain, Amaresh Mishra, Yamini Pathak, Walace Gomes-Leal, Michelle Castro da Silva, Edna Cristina Santos Franco, Celice Cordeiro de Souza, Marcelo Marques Cardoso, Carolina Ramos dos Santos, Sujata Maurya

First published in London, United Kingdom, 2021 by IntechOpen
IntechOpen is the global imprint of INTECHOPEN LIMITED, registered in England and Wales, registration number: 11086078, 5 Princes Gate Court, London, SW7 2QJ, United Kingdom
Printed in Croatia

British Library Cataloguing-in-Publication Data
A catalogue record for this book is available from the British Library

Additional hard and PDF copies can be obtained from orders@intechopen.com

Macrophages
Edited by Hridayesh Prakash
p. cm.
Print ISBN 978-1-78923-887-7
Online ISBN 978-1-78923-888-4
eBook (PDF) ISBN 978-1-78985-141-0

For EU product safety concerns:
IN TECH d.o.o., Prolaz Marije Krucifikse Kozulić 3, 51000 Rijeka, Croatia, info@intechopen.com or visit our website at intechopen.com.

Meet the editor

Dr. Hridayesh Prakash is a fellow of the Royal Society of Biology, London. Currently; he is working as an Associate Professor at the Institute of Virology and Immunology, Amity University, NOIDA. He holds expertise in macrophage immunobiology, tumor immunology/immunotherapy, cell-based immunotherapies, pulmonary infection biology, and radiation biology. The main area of his current research is to exploit various immunotherapeutics for the management of persistent bacterial/viral infections and gastric cancer. Within this frame, he is unraveling the therapeutic potential of M1 effector macrophages against solid tumors as well as various mechanisms that certain pathogens like *H. pylori, Chlamydia,* and *Mycobacteria* are exploiting for polarizing M1 effector macrophages towards M2 phenotype during chronic and persistent infections. Under this major objective, he is now validating the therapeutic impact of M1 effector macrophages for control of persistent infection driven cancer (adenocarcinoma) progression.

Contents

Preface

Macrophages are ubiquitous and an integrated part of both innate and adaptive immunity. These cells have been explored as extensive research tools in different contexts. Macrophages display a range of plasticity in their phenotype in different pathological conditions. Peripheral and tissue macrophages together constitute the reticuloendothelial system where they play a major role in sensing pathogens and tumor antigens for their effective eradication. Compared to various immune cells, macrophages display a range of plasticity that qualifies them as one of the potential target cells of the body for the clinical management of various human diseases. Due to their plastic nature, these cells are potentially involved in most immunological and physiological responses.

Several research groups, including ours, have demonstrated several MDR/XTR bacteria polarize M1 effector alveolar macrophages towards their M2 phenotype during their persistent infection. This seems to be the potential link to the sensitization for infection and maybe for development of cancer in a host

The current mandate of research in the macrophages immunobiology field mainly lies in the management of the M1/M2 imbalance to minimize the risk of cancer by chronic and persistent lung infection with intracellular pathogens like Chlamydia or Mycobacteria. This may be achieved by targeting major signaling pathways that drive the M2 phenotype and are involved in cancer development e.g. Sphingolipids, Th2/Th17 responses.

In view of the above, the major focus of this book is to discuss research methodologies, resources, and technologies from the dedicated biological community for identifying the molecular signature involved in the polarization of M1 effector macrophages to M2 during disease. Under the umbrella of this topic, the second major scope of this book is to explore how selective phenotypes of macrophages would improve existing therapies, especially on infection and cancer interface with special emphasis on lung cancers and various gastric inflammatory diseases like IBD, which are responsible for global mortality.

Hridayesh Prakash
Associate Professor,
Amity Institute of Virology and Immunology,
Amity University,
Noida, India

Section 1

Macrophage in Tumor Control

Chapter 1

Role of Macrophages in Solid Tumor Metabolism

Sibi Raj, Vaishali Chandel, Sujata Maurya and Dhruv Kumar

Abstract

Cancer cells undergo several complex processes to grow and evolve. For their survival, they manipulate the entire system and acquire the ability to gain all the energy demands from the host system itself. Tumor associated macrophages (TAMs) are macrophages abundantly present in the tumor micro environment (TME) and essentially plays a critical role in coordination with the tumor cells helping them to progress and metastasize. One of the key hallmarks in tumor cells is elevated metabolic processes such as glycolysis, fatty acid oxidation, mitochondrial oxidation, and amino acid metabolism. Macrophages help cancer cells to achieve this metabolic demand through a series of signaling events including mTOR, Akt, and PI3K pathways. The M2-like phenotype of macrophages leads to the tumorous macrophage phenotype along with the tumor cells to support tumor growth through metabolic dysregulation. Focusing upon the area of macrophage-mediated tumor metabolism in solid tumors has been a new area that provides new effective targets to treat cancer. This chapter discusses the role of macrophages in tumor metabolism and cancer progression. Targeting TAMs in tumor microenvironment through metabolic axis could be a potential therapeutic option to control the solid tumor growth and propagation.

Keywords: macrophages, hypoxia, TME, PD-1, OXPHOS, tumor microenvironment

1. Introduction

The slow pace development of solid tumors inside a human body involves a lot of complex process. It is not only the genetic mutations that play important role but also the so-called tumor microenvironment (TME), which is a silent player enhancing this process. TME has complex players such as the T-cells, dendritic cells, and macrophages in the solid tumor [1]. Among these, macrophages have three types of classification namely tumor-associated macrophages (TAMs), tissue-resident macrophages, and myeloid-derived suppressor cells (MDSCs). The most abundant tumor infiltrating immune cells in the tumor microenvironment are TAMs. These TAMs are classified into two subtypes namely M1 or M2 macrophages. Macrophages have a role in defense as well as homeostasis of cells by acquiring the capacity of phagocytosis. TAMs have reportedly been associated with several functions such as tumor initiation, progression, and metastasis with secretion of supporting factors such as cytokines, growth factors, inflammatory substrates, and proteolytic enzymes. As macrophages are known to be associated with tumor progression, understanding different signaling complexes has been an important field. The major signaling molecules involved are cytokines, growth factors, chemokines, and transforming growth factors beta, vascular endothelial growth factor, and platelet-derived growth factor. Several murine tumor models have reported TAMs as the major source for tumor-promoting factor like IL-6 [2]. The VEGF-A factor produced via TAMs specifically helps tumor cells with angiogenesis switch providing new blood vessels for tumor progression. TAMs have certain immunosuppressive functions apart from their strong inflammatory properties. Macrophages are poor producers of IL-12 but highly produce IL-10 and TGF-β with the help of STAT-3 activation [3]. The membrane-derived PDL-1 is activated on the surface of TAMs by IL-10 and TNF-A. Thus, PDL-1A has a prominent role in inhibiting the activated T-effector cells via the PD-1 receptor. TAMs are also widely reported to suppress therapeutic conditions such as chemotherapy, irradiation, and angiogenic inhibitors.

TAMs are associated with major metabolic changes associated with solid tumor progression. Macrophages can have a sudden change in their function while having a pathogen attack inside the host. The metabolic network inside a tumor cell has been a rich area of study as to decode the signaling molecules and find novel targets for the cure of cancer. Glycolysis is one such heavily activated pathway acquired by cancer cells to have sufficient energy and other key metabolites to progress and survive. TAMs highly elevate the process of glycolysis through HIF-1 stabilization and Akt/mTOR pathway [4]. Glycolysis in cancer cells also acts as intermediate for other cellular mechanisms such as the pentose phosphate pathway, TCA cycle, lipid metabolism, and amino acid metabolism. TAMs are also associated with increased OXPHOS despite having abrupt TCA cycle. Together with increased glycolytic flux and narrowed pathway of TCA cycle, OXPHOS promotes the accumulation of succinate and citrate in LPS/IFN-Y-activated macrophages. This accumulation of succinate leads to a major change in track of pathways by activating the HIF1-alpha subunit factor, which is otherwise in normal conditions inactivated by prolyl hydrolase enzymatic activity. This enhances production of pyruvate through glycolysis, which leads the macrophages to activate inflammatory cytokine production and the abrupt TCA cycle enhances the anti-microbial activity. Other metabolic functions such as the lipid metabolism are actively supported by the macrophages. These macrophages act as a source for synthesizing lipid mediators and fulfill the energy requirements in solid tumors. Macrophages utilize glycerides in lipoproteins as their major source of free fatty acids. This process is indicated by the increased production of lipoprotein lipase (LPL) in activated macrophages. M2-like macrophages have shown to have elevated consumption of amino acids in the form of glutamine as well as fatty acids [5]. An important mechanism of tumor suppression by macrophages through immunosuppressive phenotype helps solid tumors to evolve and grow. This mode of suppressed immunosurveillance in TAMs is mostly led by non-saturated fatty acid metabolism in macrophages. Mitochondrial respiration takes place with the help of lipid droplets, which regulates the catabolic process of free fatty acids (FFAs). mTOR signaling pathway has been reported to play an important role in suppressed immunosurveillance of TAMs. The mTORC1 responsibly is involved in the regulation of de novo lipid synthesis with the help of sterol-responsive element binding protein transcription factors.

Cancer cells and tumor microenvironment has a co-existing phenomenon which supports their growth and metastasis. As macrophages are one of the major immune cells actively present in the tumor microenvironment, the complex signaling procedure involved between the two is of utmost interest. CSF1 is one such major kind of cytokines that comes into play between TAMs and cancer cells to induce an immunosuppressive function to support tumor growth. The CSF-1 induction recruits the monocyte-derived macrophages toward the tumor surface and polarizes it to a M2-like phenotype, which is coupled to fatty acid oxidation [6]. This leads to the secretion of variety of immunosuppressive factors such as epidermal growth factor (EGF). Interestingly, the metabolic influence of TAMs on solid tumors is not unidirectional. Under hypoxia or increased lactate levels, TAMs secrete various cytokines associated with metabolic systems such as IL6, TNF, C-C motif chemokine ligand 5 (CCL5), and CCL18 [7]. These chemokines in particular promote metabolic processes like glycolysis as well as key glycolytic enzymes such as hexokinase-II, lactate dehydrogenase A (LDH-A), glucose-6-phosphate dehydrogenase etc. One of the major factors involved in cancer cells is anaerobic glycolysis or the famous Warburg effect. Hypoxia-inducible factor-1A (HIF-1A) is one of the key factors that activates aerobic glycolysis and thus stabilizes the long noncoding RNA from lactate-exposed TAMs to cancer cells. The main players in the immune system against tumors like the helper CD 4+ T-cells, cytotoxic CD 8+ T-cells, and natural killer (NK) cells on activation rely on elevated glycolytic metabolism, which in turn supports the tumor cells for their energy demands. On a similar note, Treg cells rely majorly on oxidative phosphorylation for bioenergetic demands. Interestingly this glucose dependency of both tumor and immune cells mediates the TAMs to limit the glycolytic flux in effector cells. This is mainly done through the expression of CD274, which is also known as PDL-1 and is an immunosuppressive molecule. Moreover, PDL-1 is upregulated in cellular types like TAMs, endothelial cells, and tumor cells due to the release of interferon gamma from effector cells [8]. This interaction delineates the immune effector functions and thus balances the metabolic competition majorly toward tumor progression.

Considering the growing knowledge on TAMs and its interaction toward solid tumors have given a green signal toward immune-based therapies to treat cancer. Majorly focusing on the delineation of M2-like macrophages or their depolarization toward M1-like phenotype TAMs. The inhibitor against CSF1R also holds a strong promise toward the treatment of such diseases. Strategies to shift the balance from M2- to M1-like phenotype macrophages are also being done using inhibitors against VEGF-A. Interestingly, considering the factor of co-interaction of TAMs with cancer cell and modulating their metabolism provide a great area to identify potential targets against these diseases. In line with this notion, inhibitors against MTORC1 surprisingly favor tumor progression as glycolysis gets inhibited in hypoxia-coupled TAMs, which ultimately favors tumor growth. The food and drug administration has approved drugs against PDL-1, which is an immune checkpoint blocker, which in turn simulates the immune system against cancer cells. In this reference, several metabolism-related antibodies can be functioned along with immune stimulators to treat certain types of cancer.

2. Macrophages

A sheer claim lead by Elie Metchnikoff stated that in "cellular (phagocytic) theory of immunity" the portion of white corpuscle holds an important significance in the elements of the immune system as well as protect the individuals from the invasion of pathogenic organisms [9]. Furthermore, macrophages show key role in immune responses and immunity, also the defensive role assigned to them is perfect depiction to execute the phagocytosis of pathogen aggregation. These are also held responsible for regulating lymphocyte activation as well as proliferation. With the help of antigens and allogenic cells, macrophages play an important role in the activation process of T- and B-lymphocytes [10]. Apart from these, macrophages also grant defense mechanism against the tumor cells, but studies conducted in the past several years describe the mechanism of tumor cell killed by macrophages [11]. Tumor-associated macrophages (TAMs) initiate and progress human cancers and angiogenesis and are important part of the tumor microenvironment. Targeting TAMs for therapeutic strategy to cure cancer is still in doubt [12]. Tumor metastasis is the parent cause of the deaths of cancer patients, adding to statement the intrinsic alterations in the tumor cells, but also implicated the cross-talk between cancer cells along with their altered components of micro-environment [12]. Tumor microenvironments (TME) are produced by TAMs, which further initiate the immune checkpoint and produce cytokines, chemokines, growth factors that are produced in T-cells. By doing this, TAMs have the most important functions in facilitating a metastatic cascade of the cancerous cells. At the same time, these trigger couple of more targets and few checkpoint blockade immunotherapies in order to oppose the tumor progression [13].

The term macrophages is generally defined as large bodies or cells that are instituted in the tissues that are present in the stationary forms. These are also regarded as the exceedingly multifaceted or the most versatile cells whose functions are based on their basic area of occupancy. Apart from this confinement, their pathophysiologic as well as physiologic contexts are considered to be very efficient in various studies [14]. Holding this significance in favor of host defense, also in primitive organisms, these tend to not only function as the recognition of the threats but at the same time engulf along with destroying the threats and in the higher organisms, such as humans. Macrophages have important roles in both immune responses whether adaptive or innate to the pathogens and also tend to serve as the mediators of inflammatory processes [15]. Macrophages are liberated as immature monocytes deriving from the bone marrow and further circulate in the blood stream in order to finally migrate into the tissues and also undergo the final differentiation into the resident macrophages that include kupffer cells in the liver, alveolar macrophages in the lung, and osteoclasts in the bone. It is a well-documented fact that macrophages have immunological and repair functions and are the first ones to arrive at the sites of wounding or infection where they carry out several functions that are assigned to them [16]. For promoting tissue repair, macrophages release proteases, growth factors, and angiogenic factors and for killing pathogens they release reactive oxygen and nitrogen radicals. They also release some chemokines or cytokines to arrange the action and recruitment of other immune cells and present the foreign antigens to cytotoxic T-cells [17]. Usually they are not lethal to cancer cells until they are triggered, for example, interferon gamma (IFN-γ) or lipopolysaccharide (LPS), but once they are triggered, the toxicity of cell is directly exerted toward tumor cells or indirectly via the secretion of factors that promote the anti-tumor functions of other cell types; thus, macrophages have pro- and anti-inflammatory properties, which depend on the signals they receive and the stage of disease they possess, that is the inflammatory balance in the microenvironment. Macrophages have multiple phenotypic expressions, which include removal of debris and tissue remodeling, antigen presentation, regulation of inflammation, target cell cytotoxicity, induction of immunity, thrombosis, and various forms of endocytosis [8].

Well promotion comes that of the tumor-associated macrophages (TAMs), which cover multiple strands of neoplastic tissues that counts in the angiogenesis as well as the vascularization, stroma formation accompanied by dissolution, and modulation that supports tumor cell growth which are a part of important enhancement and inhibition. On being activated TAMs are activated, and further gives rise to neoplastic cell death covering cytotoxicity and apoptosis, or even evokes tumor-destructive reactions led by the alteration of the tumor microvasculature. The primary lesions and metastases are known to group of solid tumors that are contented with the large numbers of tumor well associated of leukocytes. Famous as being the heterogeneous ones in the nature and consisting various as well as variable subsets of t-cells which are mainly the helpers, suppressor and cytotoxic, b-cells, these are considered to be the natural killer (nk) cells, and hence are termed macrophages. Significance of these macrophages lies in them making up to 80% of the cell mass in breast cancer patients [18]. Due to being heterogeneous in nature, macrophages possess wide range of phenotypes like M1 and M2 based on their environment stimulation. M1 phenotype is related with active microbe killing and M2 phenotype is related with tissue remodeling and angiogenesis. When these monocytes come in contact with tumor-derived anti-inflammatory molecules (i.e., IL-4, IL-10, prostaglandin E2, and transforming growth factor 1), in tumor cells they mature into M2 or polarized macrophages and produce factors that suppress T-cell proliferation and activity, possess poor antigen presenting ability, adapt scavenging for debris, repairing and remodeling of damaged and wound tissues, and promote angiogenesis [19]. In contrast to this, type I or M1 macrophages are immune effector cells that kill microorganisms and tumor cells. They present antigens and produce high levels of immune stimulatory cytokines. The M2 phenotype appears to be that which dominates in tumors, as TAMs show a similar molecular and functional profile that is characterized by low expression of differentiation-associated macrophage antigens such as carboxypeptidase M and CD51, high constitutive expression of interleukin IL-1 and IL-6, and low levels of tumor necrosis factor [20]. Tumor cells, endothelial cells, fibroblasts, and macrophages in human tumors expressed monocyte chemotactic protein (MCP). MCP and chemokines are TAMs derived from monocytes and are recruited largely by CCL2 (chemokine (C–C motif) ligand 2). MCP-1 highly is expressed in a wide range of tumor types such as meningioma, ovarian carcinoma, glioma, and squamous cell carcinoma of uterine cervix and may be the main determinant of the macrophages as suggested by some studies. Other major chemoattractants like vascular endothelial growth factor (VEGF), CCL3, CCL4, CCL5, CCL8, macrophage-colony stimulating factor (M-CSF or CSF-1), macrophage migration inhibition factor (MIF), and macrophage inflammatory protein-1 alpha(MIP-1) are involved in monocyte uptake into tumors and their levels in tumor mass often correlate positively with TAM numbers in human tumors [21].

3. Role of macrophage in tumor progression

As it is becoming clear now, the inflammatory cells survive in the tumor microenvironment and show crucial role in the development of cancer. The best example is TAMs that are important components of the mononuclear leukocyte population of solid tumors and show an indecisive association with tumors. TAMs exhibit several tumorigenesis-promoting functions, which have significant roles in the growth and progression of cancer such as these tend to qualify in providing the cytokines and also when it comes to induce tumor angiogenesis [22]. TAMs produce many types of protein digestive enzymes, growth factors, inflammatory mediators, and cytokines in tumor microenvironment that are the main factors in the metastasis of cancer cells. Not only this, TAMs' function and movement are also regulated in tumor microenvironment by cytokines and hypoxia. Some studies suggest that TAMs come in contact with cancer cells, they alter ECM and promote invasion and metastasis of cancer cell and several studies show the release of natural products by TAMs to inhibit the formation of pro-inflammatory cytokines and growth factors and also correlation with cancer metastasis and poor prognosis in various types of cancers that happen in humans [23]. The tumor in various murine models shows IL-6 (tumor-promoting) as the main source of TAMs, and also that the tumor-related myeloid cell production of IL-6 promotes proliferation in colon cells along with the apoptosis prevention through STAT3 activation. There is a Doppler effect observed in pancreatic cancer, IL-6 derived from myeloid cell initiate tumor development possess from epithelial precursor lesions through STAT3 [16]. In a specific genetic model of colorectal cancer, initiation of tumor starts with the loss of the adenomatous polyposis coli tumor suppressor gene, which results in the activation of β-catenin and further causes the barrier disruption of the epithelium. It allows the products of microbes to penetrate and moreover causes IL-23 macrophage production. In CD4+ T-cells, IL-23 drives Th17 response through IL-6 and IL-17, which initiate colorectal cancer [24].

The proportion of blood capillaries present in the non-infectious tissues mainly rests in an inactivated state in which angiogenesis transiently gets started in the perfect response in favor of certain stimuli. On the contrary, at the time of tumor initiation, an "angiogenic switch" is almost always initiated as well as turned on, which leads to vascularization of new capillaries from the inactivated state. On comparing the normal vascular network, the network of blood capillaries that are present in the tumors are basically identified by the complex and excessive branching of the blood vessels, which are contorted and also become large vessels, show irregular blood flow, microhemorrhage, and leakiness [25]. Macrophages are very particular to switch this angiogenic, that in the case of tumors mainly goes through production of vascular-endothelial growth factor A (VEGF-A) along with placental growth factor (PlGF). Talking of the specificity the blood vessels present in the tumors lacking myeloid cell-derived VEGF-A were less tortuous then having more pericyte coverage as well as the less vessel length. Above mentioned characteristics are consider successfully that show normal blood vessels. These are further counted upon modifying the bioavailability of VEGF-A in the tumors by matrix metalloproteinases processing. Adding to this fact, antibody-mediated neutralization of angiopoietin 2, the ligand for the Tie2 receptor, or macrophage depletion blocks tumor angiogenesis as well as limits tumor progression in a mouse model of breast cancer [26]. Several studies conducted on the patients having cancer in liver cells showed that the marginal macrophage density is different from the macrophage density present inside the tumor of the liver although they are directly associated along with the vascular invasion, tumor multiplicity, and also fibrous capsule formation. Furthermore, there was an important relationship observed between the density of TAMs as well as in the poor prognosis in those patients. According to Hansen et al., CD64+ macrophages (TAMs) are present in high numbers in tumor biopsies before treatment and thus show a negative relation along with clinical outcomes in the patients with the metastatic melanoma who undergo IL-2 based immunotherapy [27].

4. Tumor microenvironment

Tumor microenvironment (TME) changes continuously during the tumor development in parallel with the tumor growth. These changes on the one hand influence the immune cells' function and the complex relationship between tumor cells and these cells, and on the other hand influence its cellular content through the release of several factors, which leads to the accumulation of specific types of immune cells into the TME. Hypoxia and limitation of blood-borne nutrients are a characteristic feature of TME, while being enriched in reactive nitrogen species (RNS), protons, and other by-products released from the activated tumor cell metabolism [28, 29]. It is therefore important for the tumor cells to acclimatize their metabolism in order to survive in oxygen- and nutrients-deprived TME, and to respond to their increased demands of energy depending on their enhanced proliferation rate. The metabolic changes have been described over a century ago as "Warburg phenomenon" or "aerobic glycolysis," where tumor cells exploit glycolysis in order to provide energy regardless of the availability of oxygen [30]. Under hypoxic conditions, the cellular metabolism migrates toward anaerobic glycolysis to generate energy, rather than oxidative phosphorylation (OXPHOS), which plays a major role in terms of adenosine triphosphate (ATP) production [31]. As a consequence of the increased rate of glycolysis, the pyruvate is significantly reduced to lactate. This causes upregulated lactate levels that are released in TME by monocarboxylate transporters (MCTs) and that results in reduction in the pH levels and local acidification, while pH within the tumor remains normal [32]. In TME, this significant reduction in the pH levels causes cytotoxic environment for cells, including immune cells such as macrophages that are activated and recruited to restrict the progression of tumor and eliminate the tumor. This provides survival benefit to the cancer cells [29]. Additionally, the toxic waste, for example, lactic acid has been shown to frame and shape the functional phenotype of recruited macrophages toward more tolerogenic phenotypes and conferring them with proangiogenic and pro-tumorigenic properties [33].

5. Metabolic reprogramming of TAMs

TAMs are involved in multiple processes, which result in the promotion and progression of primary tumor facilitating metastasis. The compartment of TAM via an extensive remodeling of energy metabolism evolves over time (i.e., during treatment response and tumor progression) as well as in space (at various tumor sites) [34]. The variations in TAMs in response to the nutritional needs of solid tumor are very dynamic and TME perturbations have a major influence not only on the survival of TAM but also on tumor progression [35] (**Figure 1**).

5.1 Glucose metabolism in macrophages

TAMs majorly support the progression of tumor by (i) indirectly enhancing the nutrients' availability in the TME, (ii) providing signals to tumor cells, and (iii) mediating immunosuppressive functions. "Neoangiogenesis" is the major

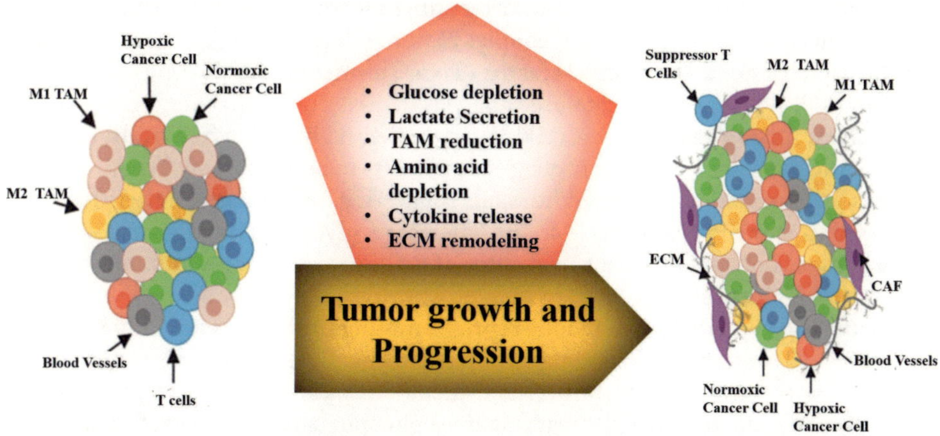

Figure 1.
TAM compartment is highly polarized toward M1-like state. Tumor progression depletes TME for the availability of glucose since it produces significant amount of lactate and polarization toward M2-like state. M2-like TAMs show restricted activity of phagocytes, chemokine, and cytokine secretion that facilitates the neoangiogenesis process, deprive TME of amino acids such as glutamine for T-cell function supporting remodulation of ECM. This metabolic alteration results in tumor growth and progression. (TAM: Tumor-associated macrophages; TME: Tumor microenvironment; ECM: Extracellular matrix).

mechanism of nutritional support to the solid tumor cells by products derived from TAM such as adrenomedullin (AMD), C-X-C motif chemokine ligand 8 (CXCL8), vascular endothelial growth factor A (VEGFA), and CXCL12 [36, 37]. Although the vasculature of tumor is functionally and phenotypically impaired, neoangiogenesis plays a crucial role for the growth of neoplasms in this scenario [38]. TME has been shown to exhibit some degree of hypoxia, which facilitates TAMs' tumor-supporting functions majorly via two mechanisms: First, hypoxic condition supports the upregulation of lipocalin 2 (LCN2), and upregulation of solute carrier family 40 member 1 (SLC40A1 or FPN). This causes the acquisition of an iron donor phenotype by TAMs, therefore enhanced availability of iron in the TME, and thus improved iron uptake by malignant cells and significant proliferation [39]. Wenes et al. investigated the character of metabolically activated hypoxic macrophage in metastasis and the blood vessel morphogenesis. It was observed that hypoxia causes TAMs to upregulate REDD1 (regulated in development and in DNA damage response 1), which causes the inhibition of mTOR, and further glycolysis inhibition. This was linked with an enhanced response to angiogenesis and leaky vessels formation. As a consequence, hypoxic TAMs migrate toward oxidative mode of metabolism coupled with decreased intake of glucose, which causes hyperactivation of endothelial cells and results in metastasis and neoangiogenesis because of the increased availability of glucose in the TME. However, the physiological relevance of such shift in humans has not been proven yet [40]. Under normoxia, TAMs exhibit downregulated activity of succinate dehydrogenase (SDH) and lower glyceraldehyde 3-phosphate dehydrogenase (GAPDH) as compared to normal macrophages, aiding their potential to function on relatively low inputs of nutrients as found in the TME. Interestingly, the activity of GAPDH was observed to be more downregulated in M2-like than M1-like macrophages infiltrating colorectal tumors in humans [41]. In a similar manner, in human gliomas TAMs derived from monocytes showed reduced glucose metabolism than tissue-resident TAMs, which was linked with poor patient survival and escalated immunosuppression in the TME. These observations in TAMs suggest that decreased glycolytic activity elicits progression of tumor via both immunosuppression and nutritional circuitries [42]. Even when a decreased glycolysis metabolism in TAMs spears to favor growth of the tumor in a majority of settings, tissue section analysis and co-culture experiments in TAMs showed that production of lactate by human medullary carcinoma cells causes a shift from OXPHOS to glycolysis, couples to upregulated secretion of interleukin 6 (IL6), lactate and tumor necrosis factor (TNF), ultimately supporting tumor progression [43]. Additionally, proteomic analysis demonstrated that enzymes involved in glycolysis such as hexokinase 2 (HK-II) are increased in TAMs from individuals with pancreatic cancer and macrophages derived from bone marrow exposed to breast carcinoma extracts from individuals suggesting an enhanced rather than decreased capacity in glycolysis [44, 45]. Therefore, glycolysis in TAMs can facilitate tumor progression and growth irrespective of an increased competition for local availability of glucose [44].

Macrophages-mediated metabolic reprogramming in tumor is not only limited to glycolysis. Through the different glycolysis intermediates, glycolysis is directly associated with various other intracellular metabolic pathways. This includes fatty acid (FA) and glutamine metabolism, pentose phosphate pathway (PPP), and amino acid metabolism.

5.2 Fatty acid and glutamine metabolism

M2-like TAMs also display increased consumption of fatty acid and glutamine. The latter represents relatively increased levels of metabolic enzymes and glutamine transporters expression, observed in both in vitro and in vivo in primary human TAMs [43]. In line with this, another study shows, glutamate ammonia ligase (GLUL) facilitates polarization of M2 by catalytic conversion of glutamate to glutamine, at least in vitro [46]. Therefore, inhibition of GLUL supports the M2-like TAMs repolarization into M1 counterparts along with enhanced flux of glycolysis and availability of succinate, suggesting the role of glutamine metabolism in TAMs regulation. Also, depletion of glutamine restrains polarization of M2-like macrophages in murine as a result of limited availability of a-ketoglutarate for epigenetic reprogramming [47]. A similar outcome ensues *N*-glycosylation inhibition suggesting the limited synthesis of aspartate-dependent UDP-N-acetyl-glucosamine (UDP-GlcNac) [47] and limited glucose-acetyl-CoA, which also plays a crucial role in epigenetic functions [48]. The former is the result of interleukin 4 (IL4)-driven activation of signal transducer and PPARG coactivator 1 beta (PPARGC1B), leading to enhanced epigenetic reprogramming and mitochondrial biogenesis toward fatty acid oxidation (FAO) [49]. Therefore, inhibition of pharmacological FAO reportedly supports repolarization of M1-like and M2-like macrophages [50], while upregulation of fatty acid synthase (FASN) in several subsets of TAM has been shown to favor pulmonary tumorigenesis because of the secretion of colony stimulating factor 1 (CSF1). In such a setting, TAMs have shown to support tumor progression by immunosuppressive cytokine interleukin 10 (IL10) release downstream of peroxisome proliferator-activated receptor delta (PPARD) [50]. The latter observation suggests the crosstalk between immune and metabolic functions in the TME. Some TAMs accumulate intracellular source of lipids in order to support metabolic fitness in tumor [51]. This suggests the alteration in majority of crucial factors involved in lipid metabolism such as monoglyceride lipase (MGLL), abhydrolase domain containing 5 (ABHD5), and acyl-CoA dehydrogenase medium chain (ACADM) [51–53]. Therefore, these observations suggest the major role of TAM metabolism on their ability to influence tumor progression and growth.

5.3 Amino acid metabolism

TAMs, exclusively pro-tumorigenic and M2-like macrophages, exhibit increased utilization of glutamine. This is linked with upregulated levels of intermediates such as uridine diphosphate N-acetylglucosamine, which are needed for N linked glycosylation of M2-like macrophages-associated receptors. Consequently, inhibiting the process of N-glycosylation and glutamine deprivation impairs polarization of M2-like macrophages along with downregulation in the TCA cycle [47]. Additionally, TAM isolated and exposed to glioblastoma cell lines exhibited enhanced gene expression related to metabolism [54]. The metabolism of L-arginine has also been shown to be associated with TAMs function. L-arginine can be used either through arginase metabolic pathway or for the synthesis of NO in macrophages. NO synthesis pathway is the characteristic feature of M1-like macrophages. Arginine is converted to L-citrulline and NO by nitric oxide synthase (iNOS). The NO furthermore downregulates OXPHOS through inhibition of electron transport chain (ETC) and TCA cycle enzymes and increases glycolysis [55, 56]. While on the other hand, expression of arginase (ARG1), enzyme that plays a crucial role in urea cycle, is the characteristic feature of M2-like macrophage, which hydrolyses arginine to urea and ornithine and restricts the availability of arginine for NO synthesis [57, 58]. TAM isolated from human ovarian and murine mammary tumors showed reduced cytotoxic properties linked with a decreased production of NO and a lower expression of iNOS in tumor-bearing mice [59, 60]. Another study demonstrated elevated expression of Arg1 in TAMs isolated from murine models. Lactate and hypoxia have been studied to be able to upregulate Arg1 expression [13]. Colegio et al. in lung cancer murine model demonstrated that Arg1fl/fl X Lysmcre/wt mice, with deficient ARG1 in macrophages, developed small-sized tumor as compared to the wild-type mice [33]. In the same study, TAMs exhibited upregulated expression of urea cycle. Additionally, metabolites such as tryptophan and cysteine derived from L-arginine are crucial mediators of myeloid-derived suppressor cells (*MDSC*). These findings highlight the role of nitrogen cycle in TAMs' function [61].

6. Signaling cross-talk between macrophages and solid tumor

Colony stimulating factor 1 (CSF1), the major cytokine, plays an important role in the interplay between TAMs and tumor cells [62]. After binding to its cognate receptor, CSF1 facilitates monocyte-derived macrophages' recruitment to tumor bed and M2-like macrophages polarization. This is accompanied with (1) upregulation of FAO [63] and (2) immunosuppressive and pro-tumorigenic factor secretion, such as IL10 [64] and epidermal growth factor (EGF) [65]. Accordingly, inhibition of colony stimulating factor 1 receptor (CSF1R) with monoclonal antibodies or small molecules supports the M1-like TAMs' accumulation [66]. This is accompanied by glycolysis restoration, mediating therapeutic effects in majority of tumor models. CSF1, VEGFA, and IL34 supporting TAMs' growth is sensitive to chemotherapeutic environmental stress, local pH, nutrient availability, and oxygen tension [33, 62]. Therefore, metabolism of lactate is exclusively relevant not only for metabolic symbiosis between normoxic and hypoxic cancer cells but also for the potential of hypoxic cancer cells to decrease TAMs toward poor M2-like glycolytic profile, exhibiting upregulation of FAO, reduced potential for antigen presentation [67]. Additionally, M2 polarization of TAMs-associated melanoma is elicited by a G-protein-coupled receptor (GPCR) signaling mechanism that senses acidification of TME induced by increased glycolysis in cancer cells [68]. Mathematical modeling accompanied with in vivo experiments revealed the potential of TAMs to support the process of neoangiogenesis. This specific metabolic alteration has additional immunological consequences, as VEGFA favors the immunosuppressive receptors' expression [69]. Upregulated activity of lactate in the TME causing hypoxic nature contributes to the arginine catabolism by arginase 1 (ARG1) and ARG2 over nitric oxide synthase 2 (NOS2), causing enhanced secretion of factors supporting tumor such as polyamines and ornithine by TAMs [33, 69]. The levels of ARG1 can be increased in M2-like TAMs by signals induced by apoptotic cancer cells [70], such as FASN-dependent pathway driven by CSF1 [42] and sphingosine-1-phosphate (S1P). Also, lactate contributes polarization of M2-like macrophages in murine breast cancer models by triggering G-protein-coupled receptor 132 (GPR132) signaling. Accordingly, upregulated levels of GPR132 elicit infiltration in breast cancer by monocyte-derived macrophages, which certainly acquire functions supporting tumor phenotype. Another receptor of lactate, hydroxycarboxylic acid receptor 1 (HCAR1), seems to be upregulated in M1-like TAMs [71]. Additionally, cancer cells' metabolic influence on TAMs is not unidirectional. Therefore, when TAMs are exposed to the hypoxic conditions or upregulated lactate levels, they secrete variety of cytokines such as TNF, IL6, C-C motif chemokine ligand 5 (CCL5), and CCL18 [72]. IL6 supports glucose metabolism by mediating 3-phosphoinositide-dependent protein kinase 1 (PDPK1) potential to phosphorylate CCL5, TNF, phosphoglycerate kinase 1 (PGK), and CCL18 enhances pro-glycolytic factors such as PGK1, HXK2, glucose-6-phosphate dehydrogenase (G6PD), lactate dehydrogenase A (LDHA), vascular cell adhesion molecule 1 (VCAM1), pyruvate dehydrogenase (PDH), pyruvate dehydrogenase kinase 1 (PDK1), and GLUT1 [73]. Apart from these findings, Warburg phenomenon is triggered, both in vitro and in vivo, by transfer of long noncoding RNA of hypoxia

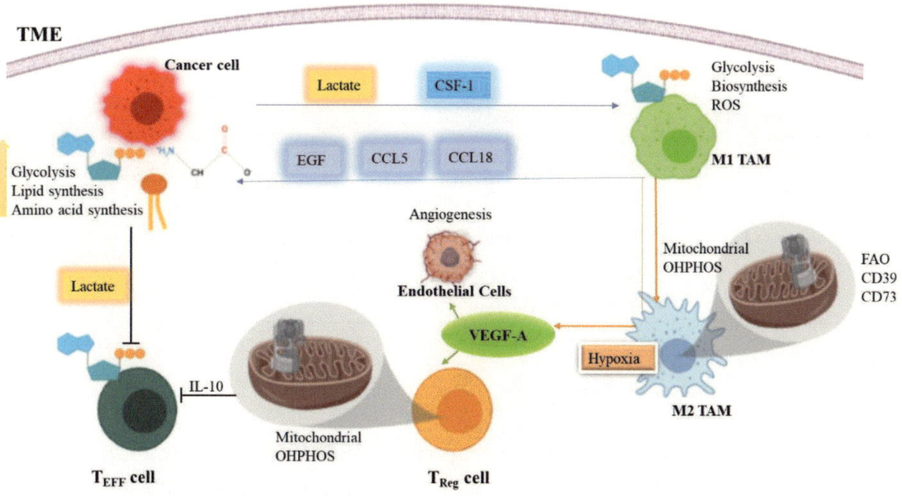

Figure 2.
Metabolic cross-talk signatures between TAMs and cancer cells. As cancer cells begin to proliferate in an uncontrollable manner, they start consuming elevated levels of glucose and other biosynthetic products to progress. This triggers the release of CSF-1 into the tumor microenvironment, which repolarizes the M1-like phenotype macrophages into M2-like macrophages. The M2-like state then releases several chemokines such as EGF, CCL-5, CCL-18, which act as immunosuppressive molecules. The expensive use of lactate by cancer cells delineates the immune function of T effector cells. Moreover, M2-like macrophages upregulated mitochondrial phosphorylation and fatty acid oxidation.

inducible factor 1 subunit alpha (HIF1A) from TAMs exposed to lactate to the tumor cells. Intriguingly, HIF1A have been shown to play a crucial role in exacerbating tumor glycolysis as well as M2 polarization. Furthermore, M2-like TAMs trigger hypoxia in an active manner [37] (**Figure 2**).

7. Therapeutic strategies against macrophages-mediated tumor metabolism

As now it is very evident that TAMs and tumors have very complex interactions between them, which supports the tumor progression, growth, and metastasis, researches across the globe are widely studying this area and finding novel targets against the immune regulators and metabolic mediators in the system. TAMs are the widely found immune cells in the tumor microenvironment and undergo complex processes to support the tumor growth. M2-like TAM phenotype is highly reported from studies to likely support the tumor progression. This conclusion supports the idea of developing antibodies that intervene with the M2 phenotype macrophage function. CCL2 blocking agents have been a promising antibody against cancer metastasis and cancer death. Therefore, strategies to deplete the M2 phenotype into non-tumorous M1 phenotype have been a promising step toward treating cancer. Drugs associated with metabolic blockers also paved a new method for treating cancer. Drugs such as inhibiting HK-II helped in the inhibition of pro-metastatic M2 phenotype of TAMs. Similarly, drugs inhibiting FAO also appear to be a promising field to target pro-tumoral macrophage as well as tumor metabolism. T-cells are the major players contributing toward immune-mediated cell metabolism, which is also approached as an effective target. PD-1 ligation with T-cells helps in the shift in metabolism from glycolysis to FAO, which maintains the longevity of T-cells and impairs their effector function (**Table 1**).

S. No.	Target	Effect	References
1	CCL2	Reduce tumor growth and metastasis in prostate and breast cancer	[74]
2.	CSF1 receptor	Antiangiogenic and antimetastasis effects in melanoma and mammary xenograft	[66]
3.	IL4Rα	Less aggressive skin tumors	[75]
4.	STAT3	Inhibited immunosuppressive cytokine profile of AAMs	[76]
5.	COX2	Suppression of breast cancer metastasis	[77]
6.	HCK	Suppression of AAM polarization, enhanced tumor immunity in colon cancer	[78]

Table 1.
Metabolic targets in tumor-associated macrophages.

Therefore, strategies that target macrophages along with tumor metabolism without encouraging the tumor cell growth have been quite challenging and have provide an effective means to treat the solid tumors.

8. Conclusion

Solid tumors are evolving over the years strongly by acquiring the capability to manipulate the host system to help them attain energy demands to survive and metastasize. One such player helping tumor cells is the macrophages. The polarization of macrophages into tumorous M2 phenotype helps the cancer cells in their glycolytic demands. Metabolic events such as amino acid metabolism and oxidative phosphorylation are triggered in cancer cells. The hypoxia event in cancer cells stimulates the signaling events toward glycolysis increasing glycolytic enzymes such as hexokinase-II, LDH-A, and pyruvate dehydrogenase. Thus, series of research has proved that macrophages has an important role to play in tumor metabolism inside the TME. This kind of immunometabolism has triggered challenges and interest in finding novel targets on this area to cure the disease cancer.

Acknowledgements

We thank our lab members for carefully reading the article and contributing valuable inputs for improving it. V.C., S.R., and D.K. were supported by the Department of Science & Technology-Science and Engineering Research Board (DST-SERB)-funded research grant (ECR/2016/001489), Govt. of India.

Author details

Sibi Raj, Vaishali Chandel, Sujata Maurya and Dhruv Kumar*
Amity Institute of Molecular Medicine and Stem Cell Research (AIMMSCR),
Amity University Noida, Uttar Pradesh, India

*Address all correspondence to: dhruvbhu@gmail.com; dkumar13@amity.edu

References

[1] Gonzalez H, Hagerling C, Werb Z. Roles of the immune system in cancer: From tumor initiation to metastatic progression. Genes and Development. 2018;**32**:1267-1284

[2] Fisher DT, Appenheimer MM, Evans SS. The two faces of IL-6 in the tumor microenvironment. Seminars in Immunology. 2014;**26**:38-47

[3] Iyer SS, Cheng G. Role of interleukin 10 transcriptional regulation in inflammation and autoimmune disease. Critical Reviews in Immunology. 2012;**32**:23-63

[4] Yu L, Chen X, Sun X, Wang L, Chen S. The glycolytic switch in tumors: How many players are involved? Journal of Cancer. 2017;**8**:3430-3440

[5] Viola A, Munari F, Sánchez-Rodríguez R, Scolaro T, Castegna A. The metabolic signature of macrophage responses. Frontiers in Immunology. 2019;**10**:1462

[6] Thapa B, Lee K. Metabolic influence on macrophage polarization and pathogenesis. BMB Reports. 2019;**52**:360-372

[7] Esquivel-Velázquez M, Ostoa-Saloma P, Palacios-Arreola MI, Nava-Castro KE, Castro JI, Morales-Montor J. The role of cytokines in breast cancer development and progression. Journal of Interferon and Cytokine Research. 2015;**35**:1-16

[8] Castro F, Cardoso AP, Gonçalves RM, Serre K, Oliveira MJ. Interferon-gamma at the crossroads of tumor immune surveillance or evasion. Frontiers in Immunology. 2018;**9**:847

[9] Dandekar RC, Kingaonkar AV, Dhabekar GS. Role of macrophages in malignancy. Annals of Maxillofacial Surgery. 2011;**1**(2):150-154

[10] Elhelu MA. The role of macrophages in immunology. Journal of the National Medical Association. 1983;**75**(3):314-317

[11] Klimpa AH, de Vries EGE, Scherphof GL, Daemenc T. A potential role of macrophage activation in the treatment of cancer. Critical Reviews in Oncology/Hematology. 2002;**44**(2):143-161

[12] Salmaninejad A, Valilou SF, Soltani A, Ahmadi S, Abarghan YJ, Rosengren RJ, et al. Tumor-associated macrophages: Role in cancer development and therapeutic implications. Cellular Oncology. 2019;**42**:591-608

[13] Lin Y, Xu J, Lan H. Tumor-associated macrophages in tumor metastasis: Biological roles and clinical therapeutic applications. Journal of Hematology & Oncology. 2019;**12**:76

[14] Mosser DM, Edwards JP. Exploring the full spectrum of macrophage activation. Nature Reviews Immunology. 2008;**8**:958-969

[15] Chaplin DD. Overview of the immune response. The Journal of Allergy and Clinical Immunology. 2010;**125**(2):S3-S23

[16] Gordon S, Martinez-Pomares L. Physiological roles of macrophages. Pflügers Archiv. 2017;**469**(3):365-374

[17] Laskin DL. Macrophages and inflammatory mediators in chemical toxicity: A battle of forces. Chemical Research in Toxicology. 2009;**22**(8):1376-1385

[18] Coussens LM, Werb Z. Inflammation and cancer. Nature. 2002;**420**(6917):860-867

[19] Gordon S, Plüddemann A. Tissue macrophages: Heterogeneity and functions. BMC Biology. 2017;**15**:53

[20] Atri C, Guerfali FZ, Laouini D. Role of human macrophage polarization in inflammation during infectious diseases. International Journal of Molecular Sciences. 2018;**19**(6):1801

[21] Deshmane SL, Kremlev S, Amini S, Sawaya BE. Monocyte Chemoattractant Protein-1 (MCP-1): An overview. Journal of Interferon & Cytokine Research. 2009;**29**(6):313-326

[22] Iijima J, Konno K, Itano N. Inflammatory alterations of the extracellular matrix in the tumor microenvironment. Cancers. 2011;**3**(3):3189-3205

[23] Bromberg J, Wang TC. Inflammation and cancer: IL-6 and STAT3 complete the link. Cancer Cell. 2009, 2009;**15**(2):79-80

[24] Testa U, Pelosi E, Castelli G. Colorectal cancer: Genetic abnormalities, tumor progression, tumor heterogeneity, clonal evolution and tumor-initiating cells. Medical Sciences. 2018;**6**(2):31

[25] Nielsen SR, Schmid MC. Macrophages as key drivers of cancer progression and metastasis. Mediators of Inflammation. 2017;**2017**:9624760

[26] Ceci C, Atzori MG, Lacal PM, Graziani G. Role of VEGFs/VEGFR-1 signaling and its inhibition in modulating tumor invasion: Experimental evidence in different metastatic cancer models. International Journal of Molecular Sciences. 2020;**21**(4):1388

[27] Ding T, Xu J, Wang F, Shi M, Zhang Y, Li S-P, et al. High tumor-infiltrating macrophage density predicts poor prognosis in patients with primary hepatocellular carcinoma after resection. Human Pathology. 2009;**40**(3):381-389

[28] Biswas SK, Mantovani A. Orchestration of metabolism by macrophages. Cell Metabolism. 2012;**15**(4):432-437

[29] Carmona-Fontaine C, Bucci V, Akkari L, Deforet M, Joyce JA, Xavier JB. Emergence of spatial structure in the tumor microenvironment due to the Warburg effect. Proceedings of the National Academy of Sciences of the United States of America. 2013;**110**(48):19402-19407

[30] Hanahan D, Weinberg RA. Hallmarks of cancer: The next generation. Cell. 2011;**144**(5):646-674

[31] Pavlova NN, Thompson CB. The emerging hallmarks of cancer metabolism. Cell Metabolism. 2016;**23**(1):27-47

[32] Pinheiro C, Sousa B, Albergaria A, et al. GLUT1 and CAIX expression profiles in breast cancer correlate with adverse prognostic factors and MCT1 overexpression. Histology and Histopathology. 2011;**10**:1279-1286

[33] Colegio OR, Chu NQ, Szabo AL, et al. Functional polarization of tumour-associated macrophages by tumour-derived lactic acid. Nature. 2014;**513**(7519):559-563

[34] Mazzone M, Menga A, Castegna A. Metabolism and TAM functions—It takes two to tango. The FEBS Journal. 2018;**285**(4):700-716

[35] Vitale I, Manic G, Coussens LM, Kroemer G, Galluzzi L. Macrophages and metabolism in the tumor microenvironment. Cell Metabolism. 2019;**30**(1):36-50

[36] Biswas SK, Allavena P, Mantovani A. Tumor-associated macrophages: Functional diversity, clinical significance, and open questions. Seminars in Immunopathology. 2013;**35**(5):585-600

[37] Chen P, Huang Y, Bong R, et al. Tumor-associated macrophages

promote angiogenesis and melanoma growth via adrenomedullin in a paracrine and autocrine manner. Clinical Cancer Research. 2011;**17**(23):7230-7239

[38] Donnem T, Reynolds AR, Kuczynski EA, et al. Non-angiogenic tumours and their influence on cancer biology. Nature Reviews. Cancer. 2018;**18**(5):323-336

[39] Mertens C, Mora J, Ören B, et al. Macrophage-derived lipocalin-2 transports iron in the tumor microenvironment. OncoImmunology. 2018;**7**(3):e1408751

[40] Wenes M, Shang M, Di Matteo M, et al. Macrophage metabolism controls tumor blood vessel morphogenesis and metastasis. Cell Metabolism. 2016;**24**(5):701-715

[41] Miller A, Nagy C, Knapp B, et al. Exploring metabolic configurations of single cells within complex tissue microenvironments. Cell Metabolism. 2017;**26**(5):788-800

[42] Müller S, Kohanbash G, Liu SJ, et al. Single-cell profiling of human gliomas reveals macrophage ontogeny as a basis for regional differences in macrophage activation in the tumor microenvironment. Genome Biology. 2017;**18**(1):234

[43] Arts RJW, Plantinga TS, Tuit S, et al. Transcriptional and metabolic reprogramming induce an inflammatory phenotype in non-medullary thyroid carcinoma-induced macrophages. OncoImmunology. 2016;**5**(12):e1229725

[44] Penny HL, Sieow JL, Adriani G, et al. Warburg metabolism in tumor-conditioned macrophages promotes metastasis in human pancreatic ductal adenocarcinoma. OncoImmunology. 2016;**5**(8):e1191731

[45] Liu D, Chang C, Lu N, et al. Comprehensive proteomics analysis reveals metabolic reprogramming of tumor-associated macrophages stimulated by the tumor microenvironment. Journal of Proteome Research. 2017;**16**(1):288-297

[46] Palmieri EM, Menga A, Martín-Pérez R, et al. Pharmacologic or genetic targeting of glutamine synthetase skews macrophages toward an M1-like phenotype and inhibits tumor metastasis. Cell Reports. 2017;**20**(7):1654-1666

[47] Jha AK, Huang SCC, Sergushichev A, et al. Network integration of parallel metabolic and transcriptional data reveals metabolic modules that regulate macrophage polarization. Immunity. 2015

[48] Hardivillé S, Hart GW. Nutrient regulation of gene expression by O-GlcNAcylation of chromatin. Current Opinion in Chemical Biology. 2016:88-94

[49] Vats D, Mukundan L, Odegaard JI, et al. Oxidative metabolism and PGC-1β attenuate macrophage-mediated inflammation. Cell Metabolism. 2006;**4**(1):13-24

[50] Hossain F, Al-Khami AA, Wyczechowska D, et al. Inhibition of fatty acid oxidation modulates immunosuppressive functions of myeloid-derived suppressor cells and enhances cancer therapies. Cancer Immunology Research. 2015;**3**(11):1236-1247

[51] Xiang W, Shi R, Kang X, et al. Monoacylglycerol lipase regulates cannabinoid receptor 2-dependent macrophage activation and cancer progression. Nature Communications. 2018;**9**(1):2574

[52] Miao H, Ou J, Peng Y, et al. Macrophage ABHD5 promotes

colorectal cancer growth by suppressing spermidine production by SRM. Nature Communications. 2016

[53] Niu Z, Shi Q, Zhang W, et al. Caspase-1 cleaves PPARγ for potentiating the pro-tumor action of TAMs. Nature Communications. 2017;**8**(1):766

[54] Choi J, Stradmann-Bellinghausen B, Savaskan N, Regnier-Vigouroux A. Human monocyte-derived macrophages exposed to glioblastoma cells and tumor-associated microglia/macrophages differ in glutamatergic gene expressions. Glia. 2015;**16**(8):1205-1213

[55] Albina JE, Mastrofrancesco B. Modulation of glucose metabolism in macrophages by products of nitric oxide synthase. American Journal of Physiology. Cell Physiology. 1993;**246** (6 pt 1)

[56] Clementi E, Brown GC, Feelisch M, Moncada S. Persistent inhibition of cell respiration by nitric oxide: Crucial role of S-nitrosylation of mitochondrial complex I and protective action of glutathione. Proceedings of the National Academy of Sciences of the United States of America. 1998;**95**(13):7631-7636

[57] Rath M, Müller I, Kropf P, Closs EI, Munder M. Metabolism via arginase or nitric oxide synthase: Two competing arginine pathways in macrophages. Frontiers in Immunology. 2014

[58] MacMicking J, Xie Q, Nathan C. Nitric oxide and macrophage function. Annual Review of Immunology. 1997:323-350

[59] DiNapoli MR, Calderon CL, Lopez DM. The altered tumoricidal capacity of macrophages isolated from tumor-bearing mice is related to reduced expression of the inducible nitric oxide synthase gene. The Journal of Experimental Medicine. 1996;**183**(4):1323-1329

[60] Klimp AH, Hollema H, Kempinga C, van der Zee AGJ, de Vries EGE, Daemen T. Expression of cyclooxygenase-2 and inducible nitric oxide synthase in human ovarian tumors and tumor-associated macrophages. Cancer Research. 2001;**61**(19):7305-7309

[61] Wu T, Sun C, Chen Z, et al. Smad3-deficient CD11b + Gr1 + myeloid-derived suppressor cells prevent allograft rejection via the nitric oxide pathway. Journal of Immunology. 2012;**189**(10):4989-5000

[62] DeNardo DG, Brennan DJ, Rexhepaj E, et al. Leukocyte complexity predicts breast cancer survival and functionally regulates response to chemotherapy. Cancer Discovery. 2011:54-67

[63] Park J, Lee SE, Hur J, et al. M-CSF from cancer cells induces fatty acid synthase and PPARβ/δ activation in tumor myeloid cells, leading to tumor progression. Cell Reports. 2015;**10**(9):1614-1625

[64] Ruffell B, Chang-Strachan D, Chan V, et al. Macrophage IL-10 blocks CD8+ T cell-dependent responses to chemotherapy by suppressing IL-12 expression in intratumoral dendritic cells. Cancer Cell. 2014;**26**(5):623-637

[65] Wyckoff J, Wang W, Lin EY, et al. A paracrine loop between tumor cells and macrophages is required for tumor cell migration in mammary tumors. Cancer Research. 2004;**64**(19):7022-7029

[66] Pyonteck SM, Akkari L, Schuhmacher AJ, et al. CSF-1R inhibition alters macrophage polarization and blocks glioma progression. Nature Medicine. 2013;**19**(10):1264-1272

[67] Allen E, Miéville P, Warren CM, et al. Metabolic symbiosis enables adaptive resistance to anti-angiogenic therapy that is dependent on mTOR signaling. Cell Reports. 2016;**15**(6):1144-1160

[68] Bohn T, Rapp S, Luther N, et al. Tumor immunoevasion via acidosis-dependent induction of regulatory tumor-associated macrophages. Nature Immunology. 2018;**19**(12):1319-1329

[69] Carmona-Fontaine C, Deforet M, Akkari L, Thompson CB, Joyce JA, Xavier JB. Metabolic origins of spatial organization in the tumor microenvironment. Proceedings of the National Academy of Sciences of the United States of America. 2017;**114**(11):2934-2939

[70] Galluzzi L, Vitale I, Aaronson SA, et al. Molecular mechanisms of cell death: Recommendations of the nomenclature committee on cell death. Cell Death and Differentiation. 2018;**25**(3):486-541

[71] Chen P, Zuo H, Xiong H, et al. Gpr132 sensing of lactate mediates tumor-macrophage interplay to promote breast cancer metastasis. Proceedings of the National Academy of Sciences of the United States of America. 2017;**114**(3):580-585

[72] Jeong H, Kim S, Hong BJ, et al. Tumor-associated macrophages enhance tumor hypoxia and aerobic glycolysis. Cancer Research. 2019;**79**(4):795-806

[73] Zhang M, Di Martino JS, Bowman RL, et al. Adipocyte-derived lipids mediate melanoma progression via FATP proteins. Cancer Discovery. 2018;**8**(8):1006-1025

[74] Qian B-Z, Li J, Zhang H, Kitamura T, Zhang J, Campion LR, et al. CCL2 recruits inflammatory monocytes to facilitate breast-tumour metastasis. Nature. 2011;**475**:222-225

[75] Linde N, Lederle W, Depner S, Rooijen NV, Gutschalk CM, Mueller MM. Vascular endothelial growth factor-induced skin carcinogenesis depends on recruitment and alternative activation of macrophages. The Journal of Pathology. 2012;**227**:17-28

[76] Edwards J, Emens L. The multikinase inhibitor sorafenib reverses the suppression of IL-12 and enhancement of IL-10 by PGE(2) in murine macrophages. International Immunopharmacology. 2010;**10**:1220-1228

[77] Na Y-R, Yoon Y-N, Son D-I, Seok S-H. Cyclooxygenase-2 inhibition blocks M2 macrophage differentiation and suppresses metastasis in murine breast cancer model. PLoS One. 2013;**8**:e63451

[78] Poh AR, Love CG, Masson F, Preaudet A, Tsui C, Whitehead L, et al. Inhibition of hematopoietic cell kinase activity suppresses myeloid cell-mediated colon cancer progression. Cancer Cell. 2017;**31**:563

Chapter 2

Targeting Tumor-Associated Macrophages by Plant Compounds

Alice Grigore

Abstract

Macrophages play an important role in cancer development, as they represent almost half of the cells forming the tumor microenvironment. They are called tumor-associated macrophages (TAMs) and most of them are alternative activated macrophages (M2 polarized), promoting cancer progression, angiogenesis and local immunosuppression. Blocking the macrophages recruitment, preventing their activation or turning M2 cells toward M1 phenotype (classic activated macrophage promoting an efficient immune response) is a modern immunotherapeutic approach for fighting cancer. Several studies showed that plant compounds (phenolics, triterpenes, coumarins, etc.) exert antitumor properties, not only by a direct toxical effect to malignant cells but also by influencing macrophage phenotypic differentiation.

Keywords: macrophage polarization, phenolic compounds, saponins, polysaccharides, coumarins, anthraquinones, alkaloids, tumor microenvironment

1. Introduction

Macrophages represent up to 50% of the cells infiltrating into the tumor microenvironment (TME) and modulation of macrophage polarization is an interesting and novel therapeutic approach in preclinical or clinical cancer research.

An increasing number of studies have also shown that tumor-associated macrophages (TAMs) can antagonize, augment or mediate the antitumor effects of cytotoxic agents, tumor irradiation, anti-angiogenic/vascular damaging agents and checkpoint inhibitors [1].

In the tumor microenvironment, TAMs are one of the major contributors in *angiogenesis* by secreting pro-angiogenic factors, such as vascular endothelial growth factor (VEGF), adrenomedullin (ADM), platelet-derived growth factor (PDGF), tumor growth factor-beta (TGF-β) and matrix metalloproteinases (MMPs). Also, TAMs promote tumor cell *invasion and metastasis* by modifying the composition of extracellular matrix and cell-cell junctions and promoting basal membrane disruption. It was demonstrated that macrophages *facilitate the metastasis* by enhancing the ability of cancer cells to enter a local blood vessel and also are involved in *immunosuppression* by inhibiting the T-cell response or by secreting immunosuppressive cytokines and proteases such as IL-10, TGF-β, arginase-1 and prostaglandins, which inhibit T-cell activation and proliferation [2].

TAMs often exhibit an array of activation states. In general, they are skewed away from the “classically” activated, tumoricidal phenotype (sometimes referred to as M1) toward an “alternatively” activated tumor-promoting one (M2) [1]. The classically activated M1 macrophages are stimulated by microbial substrates such as lipopolysaccharide, Toll-like receptor ligands and cytokines such as IFN-γ. They are characterized by secretion of pro-inflammatory cytokines such as interleukins IL-6, IL-12, IL-23 and TNF-α and express high levels of major histocompatibility complex class II (MHC-II), CD68, and CD80 and CD86 costimulatory molecules. The alternatively activated M2 macrophages are stimulated by IL-4 and IL-13, secrete IL-10 and TGF-β and express low levels of MHC-II and feature expression of CD163 and CD206 [3].

Unfortunately, M2 cells are the most representative cells of the TAM population within the tumor promoting genetic instability, local immunosuppression and stem cell nurturing [4] and providing essential support for a malignant phenotype [5].

In the early stages of cancers of the lung, colon and stomach, the macrophages in the normoxic milieu display an M1 phenotype and are associated with good prognosis, but within avascular areas of the tumor, TAMs alter the gene expression profile, favoring a protumor M2 phenotype, correlated with a bad prognosis [6]. In **Table 1** are showed recent conclusions concerning the correlation between TAMs and clinical prognostics in several tumor types. In human breast carcinomas, high TAM density is also associated with poor prognosis [7]. TAMs in renal cell carcinoma show a mixed M1/M2 phenotype. CD68 alone has a poor predictive value, while low CD11+ and high CD206+ as single variables correlated with reduced survival [8]. There is strong evidence for an inverse relationship between TAM density and clinical prognosis in solid tumors of the breast, prostate, ovary and cervix. Type I and II endometrial carcinomas had significantly higher macrophage density in both epithelial and stromal compartments than benign endometrium [9]. Type II cancers have nearly twice the TAM density of type 1 cancers and this difference may be due to M1 macrophage predominance in the stroma of type II cancers [10].

TAMs’ distribution pattern could be an independent prognostic factor for the overall survival of gastric cancer patients, invasive front-/stroma-dominant pattern having worse outcomes [11]. Studies have shown that the amount of TAMs in tumor stroma predicts the size, stage and metastasis of the gastric tumor [12]. In lung cancer, M2 subset and TAMs in tumor stroma were associated with worse survival, while M1 subset and TAMs in tumor islet were associated with favorable survival of lung cancer [13].

While most cancer research has focused upon these changes and most therapeutics are directed against these tumor cells, it is now apparent that the non-malignant cells in the microenvironment evolve along with the tumor and provide essential support for their malignant phenotype [5]. The knowledge of TAM activation status may allow the therapeutic targeting of TAMs, once TAMs’ targeting/modulating agents pass clinical trials and become widely available [6, 14]. The role of macrophages in tumor progression remains to be fully elucidated, in part due to the contrasting roles they play depending on their polarization [15]. Both the systemic and local environments play a tumor-initiating role through the generation of persistent inflammatory responses to a variety of stimuli [16]. To support this correlative data between macrophage-mediated inflammation and cancer induction, genetic ablation of the anti-inflammatory transcription factor STAT3 in macrophages results in a chronic inflammatory response in the colon that is sufficient to induce invasive adenocarcinoma. However, it is unclear whether macrophages in some inflammatory situations can kill aberrant cells before they become tumorigenic and thus be antitumoral [17].

Cancer type	TAMs as prognostic factors	Reference
Breast	CD68 as a biomarker for TAMs to evaluate the risk is better than CD163 or CD206 alone; high infiltration of TAMs was significantly associated with negative hormone receptor status and malignant phenotype	[18]
Gastric	The amount of TAMs in tumor stroma predicts the size, stage and metastasis of the gastric tumor Invasive front-/stroma-dominant pattern having worse outcomes Although CD68+ TAMs infiltration has the neutral prognostic effects on OS, the M1/M2 polarization of TAMs are predicative factors of prognosis in gastric cancer patients	[11, 12, 19]
Lung	The prognostic value of tumor-infiltrating TAMs in lung cancer is still controversial. M2 subset and TAMs in tumor stroma were associated with worse survival, while M1 subset and TAMs in tumor islet were associated with favorable survival of lung cancer. CD204-positive TAMs are the preferable marker for prognostic prediction in NSCLC Although the density of total CD68+ TAMs is not associated with overall survival, the localization and M1/M2 polarization of TAMs are potential prognostic predictors of NSCLC	[13, 20, 21]
Cervix	Tumor-infiltrating CD204+ M2 macrophages may predict poor prognosis in patients with cervical adenocarcinoma	[22]
Ovarian	CD163+ TAM infiltration was associated with poor prognosis of ovarian cancer and high M1/M2 macrophage ratio in tumor tissues predicted better prognosis	[23]
Pancreatic	Although TAM populations in tumor stroma are high, marking them as a probable prognostic factor, the multiple roles that TAMs play in pancreatic cancer progression have not yet been delineated. Additional mechanistic insight into the pathways that regulate the differentiation of TAMs from monocytes is required The density of TAMs has an impact on the overall survival of pancreatic cancer patients. M2-TAMs can be recognized as a prognostic indicator in pancreatic cancer	[24, 25]
Renal	CD68 alone has a poor predictive value, while low CD11+ and high CD206+ as single variables correlated with reduced survival	[8]
Glioblastoma	TAM, accounting for approximately 30% of the GBM bulk cell population, may explain, at least in part, the immunosuppressive features of GBMs	[26]
Hepatocellular carcinoma	The prognostic value of TAMs in patients with hepatocellular carcinoma (HCC) is still controversial. TAMs could serve as independent predictive indicators and therapeutic targets for HCC. Further trials are needed to elucidate the exact relationship and the underlying mechanism	[27]
Melanoma	Independent of their intratumoral distribution, the prevalent accumulation of M2 TAMs in MM is statistically confirmed to be a poor indicator of patients' outcome	[28]
Non-Hodgkin's lymphoma	High-density CD68+ and CD163+ TAMs, and also high CD163+/CD68+ TAMs ratio is significantly correlated with poor overall survival	[29]
Hodgkin's lymphoma	High density of either CD68+ or CD163+ TAMs is a robust predictor of adverse outcomes in adult cHL	[30]
Colorectal (CRC)	The role of tumor-associated macrophages (TAMs) in predicting the prognosis of CRC remains controversial. Still, high-density CD68+ macrophage infiltration can be a good prognostic marker	[31]
Squamous cell carcinoma of the head and neck (SCCHN)	CD68+ marker has no prognostic utility in patients with SCCHN; the M2-like marker CD163+ predicts poor prognosis	[32]

Table 1.
TAMs as potential predictive indicators in several tumor types.

Targeting a single signaling axis that promotes the immunosuppressive and protumoral functions of macrophages is inadequate as there are multiple signals involved in the communication between tumor cells and TAMs. Identifying and inhibiting key driver pathways, which are critical for both cancer cell survival and TAM activation, may offer therapeutic advantages as they disrupt the vicious positive feedback loop between tumor and TAMs [33]. Prevention of TAM accumulation and reduction of TAM presence by depleting existing TAMs represent novel strategies for an indirect cancer therapy specifically aimed at tumor-promoting cells within the microenvironment, but the challenge with this approach is to find ways for local administration of such drugs to the tumor [15]. Targeting TAM polarity toward an M1 phenotype also became a real immunotherapeutical approach in cancer, recalling responses from both innate and adaptive immune systems, leading to tumor regression [4].

Triple combination of anti-CTLA-4, anti-PD-1 and G47Δ-mIL12 was associated with macrophage influx and M1-like polarization in two glioma models [34]. A combination of a bivalent ganglioside and β-glucan, a yeast-derived polysaccharide, able to differentiate TAMs into an M1 phenotype is currently under investigation in a phase I clinical trial of patients with neuroblastoma [35]. Vadimezan, a fused tricyclic analog of flavone acetic acid, was found to repolarize macrophages in M1 phenotype, and it has been the subject of numerous preclinical studies and clinical trials [36]. Zoledronic acid, a clinical drug for cancer therapy, has been found to inhibit spontaneous mammary carcinogenesis by reverting macrophages from the M2 phenotype to the M1 phenotype [37].

2. Herbal compounds in TAM modulation

Research to date suggests that, despite the potency of cytotoxic anticancer agents and the high specificity that can be achieved by immunotherapy, neither of these two types of treatment is sufficient to eradicate the disease. Moreover, even in standard chemotherapy, there has been efficiency through the introduction into current practice of treatments with combinations of drugs [38]. In general, literature data show that the combination of conventional treatment with natural compounds exerts an additive effect caused by the alternative activation of signaling pathways that induce cell death or increase the activity of the chemotherapeutic agent. The involvement of these natural compounds (alone or in combination therapy) in the immunobiology of cancer is a branch that has not yet been studied but offers major therapeutic opportunities. Herbal compounds have many regulatory effects on macrophage polarization, but the specific mechanisms, signaling pathways and target genes involved remain incompletely understood [39]. Their effects, according to recent research studies, are summarized in **Figure 1**.

Although natural products have historically been a critical source for therapeutic drugs, sometimes natural molecules may suffer from insufficient efficacy, unacceptable pharmacokinetic properties, undesirable toxicity or reduced availability, which impedes their direct therapeutic application. Poor availability of some natural compounds, despite their pharmacological effects, limits their clinical application. In recent years, there has been an increased interest in developing nanoformulations with increased bioavailability and fewer side effects. For instance, TAM-rich tumors, due to their enhanced permeability, demonstrated an elevated retention (>700%) of the nanotherapeutic (poly(D,L-lactic-*co*-glycolic acid)-*b*-poly(ethylene glycol) (PLGA-PEG)), as compared to TAM-deficient tumors [14].

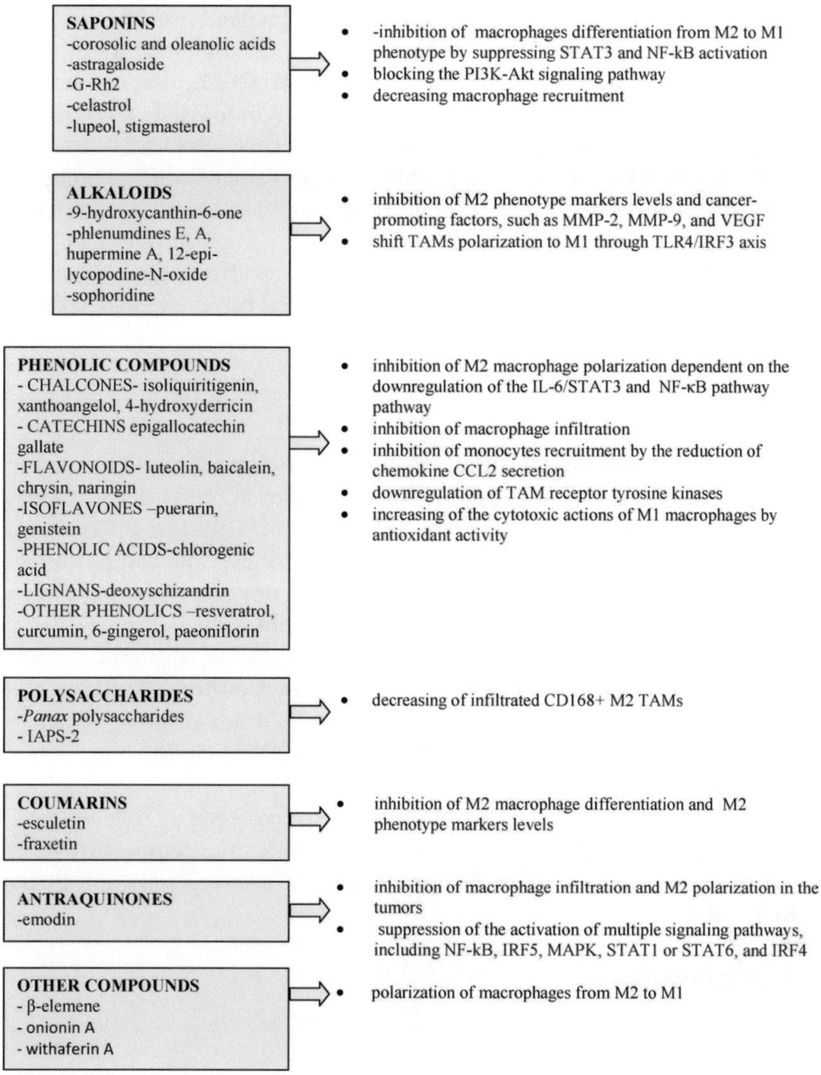

Figure 1.
Herbal compounds and their main actions on TAMs in cancer progression.

2.1 Saponins

Triterpenic compounds, including corosolic acid, tigogenin, timosaponin AIII, neoaspidistrin and oleanolic acid, suppress the CD163 expression. Corosolic and oleanolic acids change M2 polarization to M1 polarization in human monocyte-derived macrophages (HMDMs) by suppressing STAT3 and NF-kB activation. The effects of these two compounds were exerted not only on macrophages but also on glioblastoma cells, suppressing tumor cell proliferation and sensitizing tumor cells to anticancer drugs [40, 41].

M2 polarization was switched also by astragaloside IV (AS-IV, 3-O-β-D-xylopyranosyl-6-O-β-D-glucopyranosyl cycloastragenol), a natural saponin extracted from *Astragali* radix, by modulating the AMPK signaling pathway. In the intravenous lung cancer model, AS-IV treatment did not alter the percentage of macrophages but did significantly reduce the number of M2 macrophages [42]. In another study, G-Rh2, a monomeric compound extracted from *Panax ginseng* C. A.

Mey (ginseng), converts the differentiation of macrophages from M2 to M1 phenotype resulting in the decreased levels of MMPs and VEGF. By blocking the PI3K-Akt signaling pathway, the compound prevented the metastasis of lung cancer (NSCLC) cells [43]. Recently, a novel EV-liked ginseng-derived nanoparticle (GDNP) was tested in melanoma, and it altered M2 polarization both *in vitro* and *in vivo*, depending on TLR4 and MyD88 signaling and contributing to an antitumor response [44].

A potential role of celastrol, a pentacyclic triterpenoid in antimetastasis treatment, was suggested by Yang et al. [45], which found that this compound suppresses M2-like polarization by interfering with STAT6 signaling pathway after stimulation with IL-13. An active role in decreasing macrophage recruitment and tumor angiogenesis was showed for lupeol and stigmasterol in an *in vivo* model [46].

2.2 Alkaloids

Treatment with 9-hydroxycanthin-6-one, a β-carboline alkaloid isolated from the *Ailanthus altissima* stem bark, inhibited the levels of M2 phenotype markers and some cancer-promoting factors, such as MMP-2, MMP-9 and VEGF, in macrophages educated in ovarian cancer–conditioned medium. The compound also decreased the expressions of MCP-1 and RANTES, major determinants of macrophage recruitment at tumor sites, in ovarian cancer cells [47].

A regulatory effect on macrophage differentiation during tumor development exerts phlenumdines E, A, hupermine A and 12-epi-lycopodine-N-oxide isolated from the club moss *Phlegmariurus nummulariifolius* (Blume) Ching, which exhibited an inhibitory effect on IL-10–induced expression of CD163, an M2 phenotype marker, in HMDMs [48].

Sophoridine, a bioactive alkaloid extracted from the seeds of *Sophora alopecuroides* L, was able to reshape gastric cancer immune microenvironment by shifting TAM polarization to M1 and suppressing M2-TAM polarization through TLR4/IRF3 axis [49].

2.3 Phenolic compounds

2.3.1 Chalcones

In a model of azoxymethane (AOM)/dextran sodium sulfate (DSS)-induced colitis-associated tumorigenesis, it was showed that isoliquiritigenin (6′-deoxychalcone) inhibits M2 macrophage polarization depending on the downregulation of the IL-6/STAT3 pathway [50]. The same mechanism was proposed by Sumiyoshi et al. [51], for xanthoangelol and 4-hydroxyderricin, chalcones isolated from *Angelica keiskei* roots. In the *in vivo* study, the antitumor action of xanthoangelol was higher than that of 4-hydroxyderricin and it was proposed that the presence of a 4-free phenolic OH and/or the presence of a longer isoprene moiety in C-3 could be the cause of better activity of xanthoangelol. Reducing breast cancer cells' migration with the aid of M2 macrophages was achieved *in vitro* by the total flavonoid from *Glycyrrhizae Radix et Rhizoma* and isoliquiritigenin. These compounds inhibited gene and protein expression of Arg-1, upregulated gene of HO-1 and protein expression of iNOS, and enhanced the expression of microRNA 155 and its target gene SHIP1 [52].

2.3.2 Catechins

Macrophage infiltration and differentiation of macrophages into tumor-promoting M2 macrophage were decreased by epigallocatechin gallate (EGCG) treatment in murine tumor models and the molecular mechanism proposed was the downregulation of NF-κB pathway [53, 54]. EGCG can be rapidly degraded *in vivo* limiting its clinical application. A peracetate-protected EGCG (Pro-EGCG) synthesized by modification of the reactive hydroxyl groups with peracetate groups proved six times more stability than EGCG and showed greater efficacy in induction of cell death in leukemic cells. Treatment with Pro-EGCG inhibits differentiation of macrophages toward TAMs through decreasing CXCL12 expression in endometrial stromal cells with no influence on the expression level of CD163 and CD206 [55].

2.3.3 Flavonoids

Luteolin, 3 0,4 0,5,7-tetrahydroxyflavone, is a common flavonoid derived from various plants and inhibits IL-4–induced phosphorylation of STAT6 and the TAM phenotype, ameliorating the recruitment of monocytes and the migration of lung cancer cells by the reduction of chemokine CCL2 secretion from macrophages [56]. The antitumor mechanism of luteolin in non-small cell lung carcinoma (NSCLC) was mediated by downregulation of TAM receptor tyrosine kinases (RTKs), and it was found to decrease the protein levels of all three TAM RTKs in the A549 and A549/CisR cells in a dose-dependent manner [57]. In an *in vitro* tumor model, cobalt chloride (CoCl2) was used to simulate hypoxia and it was showed that luteolin decreased the expression of VEGF and MMP-9, which promote angiogenesis. In addition, luteolin also suppressed the activation of HIF-1 and phosphorylated-signal transducer and activator of STAT3 signaling, particularly within the M2-like TAMs [58].

The regulation of M2 macrophage repolarization through inhibiting PI3K/Akt signal pathway is the mechanism proposed for baicalein (5,6,7-trihydroxyflavone), a widely used Chinese herbal medicine derived from the root of *Scutellaria baicalensis*. Changing the phenotype of macrophages from M2 to M1 was supported by decreasing of M2-specific marker CD206 correlated to the increased M1-specific marker CD86. Still, the authors of the study suggested that the cytotoxic effect of baicalein on breast cancer cells directly is more pronounced than on TAMs (IC 50 of baicalein for MDA-MB-231 at 24 h, 48 h and 72 h was 79.12/50.10/34.77 μmol/L, for MCF-7 at 24 h, 48 h and 72 h was 49.76/43.73/39.44 μmol/L, for TAM at 24 h, 48 h and 72 h was 191.5/107.1/41.78 μmol/L, respectively) [59].

It has been reported that a novel chrysin (5,7-dihydroxyflavone) analog 8-bromo-7-methoxychrysin has anticancer activities with more potent bioactivity than the lead compound [60]. It also has the capacity to regulate the tumor microenvironment by inhibition of NF-κB activation, suppressing significantly the expression of the M2 macrophage marker CD163 and modulating the secretion profile of TAM cytokines [61].

According to traditional Chinese medicine (TCM) theory, herbs with Qi-tonifying character are involved in improving the defense capacity of immune system. Total flavonoids from *Glycyrrhizae Radix et Rhizoma* significantly inhibited the expression of Arg-1 (above 90% at 100 μg/mL), one of the phenotype markers of M2 macrophages, and suppressed M2 polarization of macrophages partly by inactivating STAT6 pathway. The regulation of M1 and M2 markers' expressions was partly due to the enhancement of miR-155 levels [62].

Naringin (4′,5,7 trihydroxyflavanone-7-rhamnoglucoside) exert a potential inhibitory effect on tumor progression by inducing CD169-positive and M1-like macrophages, potentially correlating with cytotoxic T-cell activation [63].

2.3.4 Isoflavones

Puerarin [4H-1-benzopyran-4-one, 8-β-D-glucopyranosyl-7-hydroxy-3-(4-hydroxyphenyl)] is the major bioactive ingredient isolated from the root of traditional Chinese medicine Ge-gen (*Radix Puerariae*) able to suppress the cell invasion and migration probably through inactivating MEK/ERK 1/2 pathway in a model of NSCLC. Also, it was showed that puerarin acts directly on macrophages by increasing M1 macrophage markers (CD197+, iNOS+ and CD40+) and reducing the expression of M2 markers (CD206+, Arg-1+ and CD163+) [64].

Another isoflavone, genistein, can inhibit the increased M2 polarization of macrophages and stemness of ovarian cancer cells by co-culture of macrophages with ovarian cancer stem-like cells through disrupting IL-8/STAT3 signaling axis [65].

2.3.5 Phenolic acids

Chlorogenic acid (5-caffeoylquinic acid, CA), the ester of caffeic acid, is a phenolic compound widely found in plants. It was showed that this compound inhibits growth of G422 glioma *in vivo*, an effect associated with a decrease of M2-like TAMs and recruitment of M1-like TAMs into tumor tissue. Low dose (1 μM) of CA could significantly inhibit the M2 macrophage-induced proliferation of glioma and breast cancer cells, mainly via STAT1 and STAT6 signaling pathways [66]. Oršolić et al. [67] concluded that the antitumor activity of CA is the result of the synergistic activities of different mechanisms by which CA acts on proliferation, angiogenesis, immunomodulation and survival. Mice with Ehrlich ascites tumor (EAT) and treated for 10 days with CA in a dose of 40 and/or 80 mg kg^{-1} showed an increase of the cytotoxic actions of M1 macrophages and inhibition of the tumor growth, probably mediated through its antioxidative activity.

2.3.6 Lignans

Deoxyschizandrin, a major dibenzocyclooctadiene lignan present in *Schisandra chinensis* berries, significantly suppressed CD163 and CD209 expression, inhibiting protumor mediator production as well as M2 polarization in TAM macrophages stimulated by the conditioned medium of A2780 cells [68].

2.3.7 Other phenolic compounds

Several studies focused on a stilbene derivative, resveratrol (3,4′,5-trihydroxystilbene), a widely studied compound that exhibits potent preventive effects on lifestyle-related disorders such as hyperlipidemia, obesity, coronary heart disease and cancer, as well as on aging. In lung cancer tumors, resveratrol induced their sluggish growth by decreasing F4/80 positive expressing cells and M2 polarization (lower expression of M2 markers-IL-10, Arg-1 and CD206), probably by STAT3 suppression [69]. Antitumor and antimetastatic effects of resveratrol (25 and 50 μM) based on the regulation of M2 macrophage activation and differentiation were confirmed by Kimura and Sumiyoshi [70], which also conducted a study for correlation of stilbene structure with biological activity. Among the nine stilbenes examined, 2,3-,3,4-, and 4,4′-dihydroxystilbene inhibited the production of MCP-1 in M2-polarized THP-1 macrophages at a concentration of 50 μM, demonstrating that the inhibitory effects of stilbenes with dihydroxy groups on the production of MCP-1 were greater than those with mono-hydroxyl groups. Dihydroxystilbene at 25 and 50 μM, 3,4-dihydroxystilbene at 50 μM, and 4,4′-dihydroxystilbene at 10, 25 and 50 μM significantly inhibited the production of IL-10 by M2 THP-1 macrophages. The three dihydroxystilbenes, 2,3-, 3,4-, and 4,4′-dihydroxystilbenes, at concentrations of 10–50 μM inhibited p-STAT3 increase during M2 THP-1 macrophage differentiation induced by IL-4 plus IL-13 [71].

The resveratrol analogue, HS-1793 (4-(6-hydroxy-2-naphthyl)-1,3-benzenediol), was also shown to elevate the level of IFN-γ production conducting reprograming of TAMs M2 phenotype [72].

Curcumin ((1*E*,6*E*)-1,7-Bis(4-hydroxy-3-methoxyphenyl)hepta-1,6-diene-3,5-dione), a natural phenol and the main active ingredient in turmeric, acts in several ways as a suppressor of macrophage functions. Even though curcumin has previously received considerable attention from researchers as an anti-inflammatory agent, it has a promising future in the area of immunomodulation [73]. Most of the studies on curcumin focused on the anti-inflammatory effect, promoting the conversion of macrophages from M1 to an anti-inflammatory and protective M2 phenotype [73]. Gao et al. [74] demonstrated that curcumin plays a key role in M2 polarization in two ways: (1) via the inhibition of DNA methyltransferase3b (DNMT3b), overexpression of which can promote increased M1 polarization, and (2) via increased phosphorylation of signal transducer and activator of transcription STAT-6, an important transcription factor activated by IL-4 and IL-10. Other studies showed that curcumin also induces TAMs re-polarization from tumor-promoting M2 phenotype toward the more antitumor M1 phenotype in tumor-bearing hosts, mediated by inhibition of STAT3 activity [75]. Curcumin administration and delivery to glioblastoma brain tumors (GBM) caused a dramatic re-polarization of TAMs from an M2 to M1 phenotype and tumor remission in 50–60% of GBM-bearing mice [76]. Hydrazinocurcumin, a synthetic analog of curcumin encapsulated within nanoparticles, reeducates TAMs to an M1-like phenotype IL-10 low IL-12 high TGF-β low [54].

It was showed that TriCurin, a synergistic formulation of curcumin, resveratrol, and epicatechin gallate (molar ratio C:E:R: 4:1:12.5) can shift TAM polarity in HPV-positive HNSCC by silencing the M2 TAM and activating/recruiting a discrete population of M1 TAM while maintaining a constant number of overall intra-tumor Iba1+ TAM, along with expression of activated STAT3 and induction of activated STAT1 and NF-kB (p65) [77]. Moreover, a liposomal formulation of TriCurin with increased bioavailability (TrLp) was able to cause repolarization of M2-like tumor (GBM)-associated microglia/macrophages to the tumoricidal M1-like phenotype and intra-GBM recruitment of activated natural killer cells [78].

In a urethane-induced lung carcinogenic model, lung carcinogenesis was ameliorated with increased M1 macrophages and decreased M2 macrophages in the lung interstitial by administration of 6-gingerol ((*S*)-5-hydroxy-1-(4-hydroxy-3-methoxyphenyl)-3-decanone), the main bioactive component in ginger (*Zingiber officinale* Roscoe). M2 macrophage-resetting efficacy of 6-gingerol was confirmed in a Lewis lung cancer allograft model and the mechanism proposed was the reduction of Arg-1 and ROS levels and elevation of L-arginine and NO levels [79].

Also, it was showed that paeoniflorin, one of the major active constituents of *Paeonia lactiflora* Pallas, inhibits the alternative activation of macrophages in subcutaneous xenograft tumors of the C57BL/6 J mice at doses of 40 and 20 $mg{\cdot}kg^{-1}$ [80].

2.4 Polysaccharides

It was suggested that modulation of TAM polarization was implicated in the antitumor immunostimulatory activity of polysaccharides from *Panax japonicus* (ginseng). The transcription and production of TGF-β and IL-10, two well-known immunosuppressive cytokines secreted by TAMs, were reduced in response to *Panax* polysaccharides and also the number of infiltrated CD168+ M2 TAMs was substantially declined although the number of CD68+ total macrophages in transplanted tumor tissues remained almost unchanged [81]. A significant inhibition of Arg-1 expression (above 90% at 100 μg/mL), one of the phenotype markers of

M2 macrophages, was also observed for the ethanol extract of Ginseng *Radix et Rhizoma* [62]. Recently, Chen et al. [82], showed that water extract of Ginseng and Astragalus could be a novel option for integrative cancer therapies due mainly to their ability to regulate macrophage polarization.

In a murine model of sarcoma, immunotherapy with IAPS-2 (acidic polysaccharide, namely IAPS-2, from the root of *Ilex asprella*) demonstrated that it could significantly inhibit the growth of tumors via modulating the function of TAMs and increase the animal survival rate [83]. Similar results were obtained with an aqueous extract of *Trametes robiniophila* Murr (Huaier), a sandy beige mushroom found on the truck of trees and has been widely used in TCM for approximately 1600 years for its antitumor, antiangiogenic and immunomodulatory effects. Huaier not only modulates the macrophage polarization but also could inhibit the macrophage-induced angiogenesis by decreasing the expression of VEGF, MMP2 and MMP9, thus inhibiting the formation of new blood vessels in tumor [84].

2.5 Coumarins

Esculetin (6,7-dihydroxycoumarin) and fraxetin (6-methoxy-7,8-dihydroxycoumarin) (50, 75 and 100 μM) inhibited the production of IL-10, MCP-1 and TGF-β-1 in macrophages and the phosphorylation of STAT 3 without affecting its expression during the differentiation of M2 macrophages. Esculetin also suppressed the increased production of these cytokines during M2 macrophage differentiation at 10–100 μM. On the other hand, daphentin (7,8-dihydroxycoumarin) had no such effects, revealing that coumarins with two hydroxyl groups at the 6 and 7 positions (esculetin) or coumarins with a methoxy group at the 6 and two hydroxyl groups at the 7 and 8 positions (fraxetin) are more active, exhibiting antitumor and antimetastatic actions in osteosarcoma LM8 cells [85]. The antitumor and antimetastatic actions of esculetin may be due to the dual actions at tumor and TAM sites: inhibition of the expression of cyclin D 1 and CDK4 in osteosarcoma LM8 cells, and also decreasing the STAT 3 phosphorylation in macrophages. In the case of fraxetin, the effects are partly attributed to the inhibition of M2 macrophage differentiation [85].

A classical formula of traditional Chinese medicine (TCM) to alleviate lung cancer–related symptoms is Bu-Fei decoction (BFD), consisting of six herbal Chinese medicines-*Codonopsis pilosula*, *Schisandra chinensis*, *Rehmannia glutinosa*, *Astragalus* sp., *Aster* sp. and *Morus* sp.-but it has not been established whether it induces an antitumor effect or it modulates the tumor microenvironment. The result of an *in vivo* study revealed that BFD successfully interrupted the interaction between tumor cells and TAMs by inhibiting the expression of two important markers: IL-10 (correlated with late stage (stage II, III and IV), lymph node metastases, pleural invasion, lymphovascular invasion and poor differentiation in NSCLC patients) and PD-L1 (correlated with poor prognosis in a number of human cancers, including breast cancer, kidney cancer and NSCLCs) [86].

2.6 Anthraquinones

It has been shown that emodin (6-methyl-1,3,8-trihydroxyanthraquinone), the active ingredient of several Chinese herbs including Rhubarb (*Rheum palmatum*), inhibits the growth of a variety of tumors and enhances the responsiveness of tumors to chemotherapy agents. In breast cancer, emodin directly inhibited macrophage infiltration and M2 polarization in the tumors, independent of tumor size [87]. Previously, Jia et al. [88], showed that emodin is not cytotoxic to breast cancer cells at concentration achieved *in vivo* (up to 30 μM) and it failed to affect macrophage infiltration in primary tumors. In contrast to its lack of effects on primary tumors, emodin dramatically suppressed lung metastasis by diminishing phosphorylation of STAT6 and C/EBPβ signaling upon IL-4 stimulation [88]. Further, it was showed that emodin suppresses the activation of multiple signaling pathways, including NF-kB, IRF5, MAPK, STAT1 or STAT6, and IRF4, depending on the environmental settings. It acts mostly on M2 polarization, suggesting that emodin could be most beneficial for patients with M2 macrophage-driven diseases [89].

2.7 Other herbal compounds/preparations

In oral squamous cell carcinoma (OSCC) animal models, highly pure super critical CO_2 leaf extract of *Azadirachta indica* (Neem) induces an M1 phenotype in TAMs *in vivo*, and the primary active component, nimbolide (a limonoid tetra-nortriterpenoid with an α,β-unsaturated ketone system and a δ-lactone ring) has significant anticancer activity in established OSCC xenografts [90]. β-Elemene, a widely known sesquiterpene, regulated the polarization of macrophages from M2 to M1, inhibiting the proliferation, migration and invasion of lung cancer cells and enhancing its radiosensitivity [91].

Onionin A (ONA), a natural low molecular weight compound containing sulfur isolated from onions, inhibited the EOC cell-induced M2 polarization of HMDMs, and STAT3 activation was significantly inhibited by ONA treatment in all cell lines [92].

Adjunctive treatment with Withaferin A, the most abundant constituent of *Withania somnifera* (Ashwagandha) root extract, reduced myeloid cell-mediated immune suppression and polarized immunity toward a tumor-rejecting type 1 phenotype, facilitating the development of antitumor immunity [93].

Traditional Chinese medicine provides pharmacologically efficient preparates such as KSG-002, a hydroalcoholic extract of radices *Astragalus membranaceus* and *Angelica gigas* at 3: 1 ratio that suppresses breast cancer growth and metastasis through targeting NF-κB–mediated TNF α production in macrophages [94] and SH003, mixed extract from *Astragalus membranaceus*, *Angelica gigas* and *Trichosanthes kirilowii* Maximowicz that suppresses highly metastatic breast cancer growth and metastasis by inhibiting STAT3-IL-6 signaling path [95].

Traditional Chinese medicine Jianpi Yangzheng Decoction (JPYZ) used for improving the quality of life and prolonging the survival of gastric cancer patients was more effective compared with Jianpi Yangzheng Xiaozheng Decoction (JPYZXZ) for inducing the phenotypic change in macrophages from M2 to M1. JPYZXZ inhibits the gastric cancer EMT more effectively than JPYZ, but JPYZ primarily works to regulate the phenotypic change in macrophages from M2 to M1 [96].

CXCL-1 was also found to be a cytokine secreted by tumor-associated macrophage, which recruits myeloid-derived suppressor cells to form pre-metastatic niche and led to liver metastasis from colorectal cancer. The current study demonstrated that after administration of XIAOPI formula (consisting of 10 herbs including *Epimedium brevicornum*, *Cistanche deserticola*, *Leonurus heterophyllus*, *Salvia miltiorrhiza*, *Curcuma aromatica*, *Rhizoma Curcumae*, *Ligustrum lucidum*, *Radix Polygoni Multiflori preparata*, *Crassostrea gigas* and *Carapax trionycis*), the density of TAMs decreased significantly and the level of CXCL-1 was also inhibited in both mouse plasma and cellular supernatants. When CXCL-1 cytokine was co-administrated with XIAOPI formula, the antimetastatic property of XIAOPI formula was blocked, indicating that CXCL-1 might be the principal gene involved in the network regulating the action of XIAOPI formula [97].

3. Conclusions

Macrophages, as key players in the tumor microenvironment, play essential roles in maintenance and progression of malignant state. Due to their plasticity, these cells balance between pro- and antitumoral effects in close correlation to specific factors. Recent immunotherapeutic strategies focus on tumor-associated macrophages in two main directions: to inhibit protumor macrophages and their suppressive effects (CCL2 inhibitors, trabectedin, zoledronic acid, JAK/STAT inhibitors, etc.) and to activate TAMs to an antitumor phenotype (TLR and CD40 agonists, PI3kδ inhibitor, VEGF and Ang2 inhibitors, etc.).

Several natural compounds/herbal extracts were studied as therapeutic/supportive agents for macrophage modulation in different types of cancers, most of them being able to change M2 polarization (protumoral) to M1 polarization (antitumoral). They belong to various classes of herbal compounds: saponins (corosolic and oleanolic acids, astragaloside, ginsenosides, celastrol, etc.), alkaloids (9-hydroxycanthin-6-one, phlenumdines E, A, hupermine A and 12-epilycopodine-N-oxide, sophoridine, etc.), flavonoids and polyphenolcarboxylic acids (isoliquiritigenin, xanthoangelol and 4-hydroxyderricin, baicalein, naringin, genistein, deoxyschizandrin, chlorogenic acid, curcumin, 6-gingerol and paeoniflorin), polysaccharides (isolated from various vegetal sources), coumarins (esculetin, fraxetin, etc.), and anthraquinones (emodin). This action is most probably achieved by downregulation of the STAT3, STAT 6 and NF-kB pathways with consecutive modulation of the secretory profile of TAM cytokines.

TCM supports the dual approach of cancer therapy, to destroy cancer cells on one hand and to improve patients' immunological status on the other hand. For several preparations such as Jianpi Yangzheng Decoction, Bu-Fei decoction and XIAOPI formula, research studies proved the correlation between cancer cells and tumor microenvironment and the effective intervention of these herbal products in delaying/breaking the tumorigenic process.

Low solubility of some herbal compounds limits their clinical application and it conducted to designing of new analogs with improved bioavailability-ginseng-derived nanoparticles, peracetate-protected EGCG, chrysin and resveratrol analogs.

By now, many herbal compounds have been shown to exhibit antitumor effects in various cancer types. Further, more researches need to be focused on the influence of these valuable compounds/preparations on modulation of the tumor microenvironment, as key element in the relation of tumor-host.

Acknowledgements

This research was financially supported by the Ministry of Research and Innovation in the frame of the project PN.16.41.01.01/2018, CORE Program.

Author details

Alice Grigore
National Institute for Chemical-Pharmaceutical Research and Development, Bucharest, Romania

*Address all correspondence to: alicearmatu@yahoo.com

References

[1] Yang M, McKay D, Pollard J, Lewis C. Diverse functions of macrophages in different tumor microenvironments. Cancer Research. 2018;**78**(19):5492-5503. DOI: 10.1158/0008-5472.CAN-18-1367

[2] Aras S, Zaidi R. TAMeless traitors: Macrophages in cancer progression and metastasis. British Journal of Cancer. 2017;**117**(11):1583-1591. DOI: 10.1038/bjc.2017.356

[3] Barros M, Hauck F, Dreyer JH, Kempkes B, Niedobitek G. Macrophage polarisation: An immunohistochemical approach for identifying M1 and M2 macrophages. PLoS One. 2013;**8**(11):e80908. DOI: 10.1371/journal.pone.0080908

[4] Najafi M, Goradel NH, Farhood B, Salehi E, Nashtaei MS, Khanlarkhani N, et al. Macrophage polarity in cancer: A review. Journal of Cellular Biochemistry. 2019;**120**(3):2756-2765. DOI: 10.1002/jcb.27646

[5] Joyce JA, Pollard JW. Microenvironmental regulation of metastasis. Nature Reviews. Cancer. 2009;**9**(4):239-252. DOI: 10.1038/nrc2618

[6] Sousa S, Régis R, Lintunen M, Kronqvist P, Sandholm J, Mönkkönen J, et al. Human breast cancer cells educate macrophages toward the M2 activation status. Breast Cancer Research. 2015;**17**:101. DOI: 10.1186/s13058-015-0621-0

[7] Choi J, Gyamfi J, Jang H, Koo JS. The role of tumor-associated macrophage in breast cancer biology. Histology and Histopathology. 2018;**33**:133-145. DOI: 10.14670/HH-11-916

[8] Kovaleva O, Samoilova D, Shitova M, Gratchev A. Tumor associated macrophages in kidney cancer. Analytical Cellular Pathology. 2016;**2016**:9307549

[9] Dun E, Hanley K, Wieser F, Bohman S, Taylor R. Infiltration of tumor-associated macrophages is increased in the epithelial and stromal compartments of endometrial carcinomas. International Journal of Gynecological Pathology. 2013;**32**(6):576-584. DOI: 10.1097/PGP.0b013e318284e198

[10] Kelly MG, Francisco AM, Cimic A, Wofford A, Fitzgerald NC, Yu J, et al. Type 2 endometrial cancer is associated with a high density of tumor-associated macrophages in the stromal compartment. Reproductive Sciences. 2015;**22**(8):948-953. DOI: 10.1177/1933719115570912

[11] Liu JY, Peng CW, Yang GF, Hu WQ, Yang XJ, Huang CQ, et al. Distribution pattern of tumor associated macrophages predicts the prognosis of gastric cancer. Oncotarget. 2017;**8**(54):92757-92769. DOI: 10.18632/oncotarget.21575

[12] Räihä M, Puolakkainen P. Tumor-associated macrophages (TAMs) as biomarkers for gastric cancer: A review. Chronic Diseases and Translational Medicine. 2018;**4**(3):156-163. DOI: 10.1016/j.cdtm.2018.07.001

[13] Wu P, Wu D, Zhao L, Huang L, Chen G, Shen G, et al. Inverse role of distinct subsets and distribution of macrophage in lung cancer prognosis: A meta-analysis. Oncotarget. 2016;**7**(26):40451-40460. DOI: 10.18632/oncotarget.9625

[14] Ovais M, Guo M, Chen C. Tailoring nanomaterials for targeting tumor-associated macrophages. Advanced Materials. 2019;**31**(19):e1808303. DOI: 10.1002/adma.201808303

[15] Poh AR, Ernst M. Targeting macrophages in cancer: From bench to bedside. Frontiers in Oncology. 2018;**8**:49. DOI: 10.3389/fonc.2018.00049

[16] Balkwill FR, Mantovani A. Cancer-related inflammation: Common themes and therapeutic opportunities. Seminars in Cancer Biology. 2012;**22**(1):33-40. DOI: 10.1016/j.semcancer.2011.12.005

[17] Noy R, Pollard J. Tumor-associated macrophages: From mechanisms to therapy. Immunity. 2014;**41**(1):49-61. DOI: 10.1016/j.immuni.2014.06.010

[18] Zhao X, Qu J, Sun Y, Wang J, Liu X, Wang F, et al. Prognostic significance of tumor-associated macrophages in breast cancer: A meta-analysis of the literature. Oncotarget. 2017;**8**(18):30576-30586. DOI: 10.18632/oncotarget.15736

[19] Yin S, Huang J, Li Z, Zhang J, Luo J, Lu C, et al. The prognostic and clinicopathological significance of tumor-associated macrophages in patients with gastric cancer: A meta-analysis. PLoS One. 2017;**12**(1):e0170042. DOI: 10.1371/journal.pone.0170042

[20] Li Z, Maeda D, Yoshida M, Umakoshi M, Nanjo H, Shiraishi K, et al. The intratumoral distribution influences the prognostic impact of CD68- and CD204-positive macrophages in non-small cell lung cancer. Lung Cancer. 2018;**123**:127-135. DOI: 10.1016/j.lungcan.2018.07.015

[21] Mei J, Xiao Z, Guo C, Pu Q, Ma L, Liu C, et al. Prognostic impact of tumor-associated macrophage infiltration in non-small cell lung cancer: A systemic review and meta-analysis. Oncotarget. 2016;**7**(23):34217-34228. DOI: 10.18632/oncotarget.9079

[22] Kawachi A, Yoshida H, Kitano S, Ino Y, Kato T, Hiraoka N. Tumor-associated CD204+ M2 macrophages are unfavorable prognostic indicators in uterine cervical adenocarcinoma. Cancer Science. 2018;**109**(3):863-870. DOI: 10.1111/cas.13476

[23] Yuan X, Zhang J, Li D, Mao Y, Mo F, Du W, et al. Prognostic significance of tumor-associated macrophages in ovarian cancer: A meta-analysis. Gynecologic Oncology. 2017;**147**(1):181-187. DOI: 10.1016/j.ygyno.2017.07.007

[24] Lankadasari M, Mukhopadhyay P, Mohammed S, Harikumar K. TAMing pancreatic cancer: Combat with a double edged sword. Molecular Cancer. 2019;**18**:48. DOI: 10.1186/s12943-019-0966-6

[25] Yu M, Guan R, Hong W, Zhou Y, Lin Y, Jin H, et al. Prognostic value of tumor-associated macrophages in pancreatic cancer: A meta-analysis. Cancer Management and Research. 2019;**11**:4041-4058. DOI: 10.2147/CMAR.S196951

[26] Morisse M, Jouannet S, Dominguez-Villar M, Sanson M, Idbaih A. Interactions between tumor-associated macrophages and tumor cells in glioblastoma: Unraveling promising targeted therapies. Expert Review of Neurotherapeutics. 2018;**18**(9):729-737. DOI: 10.1080/14737175.2018.1510321

[27] Ding W, Tan Y, Qian Y, Xue W, Wang Y, Jiang P, et al. Clinicopathologic and prognostic significance of tumor-associated macrophages in patients with hepatocellular carcinoma: A meta-analysis. PLoS One. 2019;**14**(10):e0223971. DOI: 10.1371/journal.pone.0223971

[28] Falleni M, Savi F, Tosi D, Agape E, Cerri A, Moneghini L, et al. M1 and M2 macrophages' clinicopathological significance in cutaneous melanoma. Melanoma Research. 2017;**27**(3):200-210. DOI: 10.1097/CMR.0000000000000352

[29] Xu X, Li Z, Liu J, Zhu F, Wang Z, Wang J, et al. The prognostic value of tumour-associated macrophages in non-Hodgkin's lymphoma: A systematic review and meta-analysis. Scandinavian Journal of Immunology. 2020;**91**(1):e12814. DOI: 10.1111/sji.12814

[30] Guo B, Cen H, Tan X, Ke Q. Meta-analysis of the prognostic and clinical value of tumor-associated macrophages in adult classical Hodgkin lymphoma. BMC Medicine. 2016;**14**(1):159

[31] Zhao Y, Ge X, Xu X, Yu S, Wang J, Sun L. Prognostic value and clinicopathological roles of phenotypes of tumour-associated macrophages in colorectal cancer. Journal of Cancer Research and Clinical Oncology. 2019;**145**(12):3005-3019. DOI: 10.1007/s00432-019-03041-8

[32] Troiano G, Caponio VCA, Adipietro I, Tepedino M, Santoro R, Laino L, et al. Prognostic significance of CD68⁺ and CD163⁺ tumor associated macrophages in head and neck squamous cell carcinoma: A systematic review and meta-analysis. Oral Oncology. 2019;**93**:66-75. DOI: 10.1016/j.oraloncology.2019.04.019

[33] Petty AJ, Yang Y. Tumor-associated macrophages: Implications in cancer immunotherapy. Immunotherapy. 2017;**9**(3):289-302. DOI: 10.2217/imt-2016-0135

[34] Saha D, Martuza R, Rabkin S. Macrophage polarization contributes to glioblastoma eradication by combination immunovirotherapy and immune checkpoint blockade. Cancer Cell. 2017;**2**(2):253-267.e5. DOI: 10.1016/j.ccell.2017.07.006

[35] Kushner BH, Cheung IY, Modak S, Kramer K, Ragupathi G, Cheung NK. Phase I trial of a bivalent gangliosides vaccine in combination with beta-glucan for high-risk neuroblastoma in second or later remission. Clinical Cancer Research. 2014;**20**:1375-1382. DOI: 10.1158/1078-0432.CCR-13-1012

[36] Genard G, Lucas S, Michiels C. Reprogramming of tumor-associated macrophages with anticancer therapies: Radiotherapy versus chemo- and immunotherapies. Frontiers in Immunology. 2017;**8**:828. DOI: 10.3389/fimmu.2017.00828

[37] Coscia M, Quaglino E, Iezzi M, Curcio C, Pantaleoni F, Riganti C, et al. Zoledronic acid repolarizes tumour-associated macrophages and inhibits mammary carcinogenesis by targeting the mevalonate pathway. Journal of Cellular and Molecular Medicine. 2010;**14**(12):2803-2815. DOI: 10.1111/j.1582-4934.2009.00926.x

[38] Ulrich-Merzenich G. Combination screening of synthetic drugs and plant derived natural products—Potential and challenges for drug development. Synergy. 2014;**1**(1):59-69. DOI: 10.1016/j.synres.2014.07.011

[39] Wang Y, Smith W, Hao D, He B, Kong L. M1 and M2 macrophage polarization and potentially therapeutic naturally occurring compounds. International Immunopharmacology. 2019;**70**:459-466. DOI: 10.1016/j.intimp.2019.02.050

[40] Fujiwara Y, Takeya M, Komohara Y. A novel strategy for inducing the antitumor effects of triterpenoid compounds: Blocking the protumoral functions of tumor-associated macrophages via STAT3 inhibition. BioMed Research International. 2014; Article ID 348539. DOI: 10.1155/2014/348539

[41] Fujiwara Y, Komohara Y, Ikeda T, Takeya M. Corosolic acid inhibits glioblastoma cell proliferation by suppressing the activation of signal transducer and activator of

transcription-3 and nuclear factor-kappa B in tumor cells and tumor-associated macrophages. Cancer Science. 2011;**102**(1):206-211. DOI: 10.1111/j.1349-7006.2010.01772.x

[42] Xu F, Cui Q, Wei Y, Cui J, Qiu J, Hu LL, et al. Astragaloside IV inhibits lung cancer progression and metastasis by modulating macrophage polarization through AMPK signaling. Journal of Experimental & Clinical Cancer Research. 2018;**37**(1):207. DOI: 10.1186/s13046-018-0878-0

[43] Li H, Huang N, Zhu W, Wu J, Yang X, Teng W, et al. Modulation the crosstalk between tumor-associated macrophages and non-small cell lung cancer to inhibit tumor migration and invasion by ginsenoside Rh2. BMC Cancer. 2018;**18**(1):579. DOI: 10.1186/s12885-018-4299-4

[44] Cao M, Yan H, Han X, Weng L, Wei Q, Sun X, et al. Ginseng-derived nanoparticles alter macrophage polarization to inhibit melanoma growth. Journal for Immunotherapy of Cancer. 2019;**7**(1):326. DOI: 10.1186/s40425-019-0817-4

[45] Yang Y, Cheng S, Liang G, Honggang L, Wu H. Celastrol inhibits cancer metastasis by suppressing M2-like polarization of macrophages. Biochemical and Biophysical Research Communications. 2018;**503**(2):414-419. DOI: 10.1016/j.bbrc.2018.03.224

[46] Kangsamaksin T, Chaithongyot S, Wootthichairangsan C, Hanchaina R, Tangshewinsirikul C, Svasti J. Lupeol and stigmasterol suppress tumor angiogenesis and inhibit cholangiocarcinoma growth in mice via downregulation of tumor necrosis factor-α. PLoS One. 2017;**12**(12):e0189628. DOI: 10.1371/journal.pone.0189628

[47] Jeong M, Kim HM, Ahn JH, Lee KT, Jang DS, Choi JH. 9-Hydroxycanthin-6-one isolated from stem bark of Ailanthus altissima induces ovarian cancer cell apoptosis and inhibits the activation of tumor-associated macrophages. Chemico-Biological Interactions. 2018;**280**:99-108. DOI: 10.1016/j.cbi.2017.12.011

[48] Nakayama W, Fujiwara Y, Kosuge Y, Monthakantirat O, Fujikawa K, Watthana S, et al. Phlenumdines D and E, new *Lycopodium* alkaloids from *Phlegmariurus nummulariifolius*, and their regulatory effects on macrophage differentiation during tumor development. Phytochemistry Letters. 2019;**29**:98-103. DOI: 10.1016/j.phytol.2018.11.010

[49] Zhuang H, Dai X, Zhang X, Mao Z, Huang H. Sophoridine suppresses macrophage-mediated immuno-suppression through TLR4/IRF3 pathway and subsequently upregulates CD8+ T cytotoxic function against gastric cancer. Biomedicine & Pharmacotherapy. 2020;**121**:109636. DOI: 10.1016/j.biopha.2019.109636

[50] Zhao H, Zhang X, Chen X, Li Y, Ke Z, Tang T, et al. Isoliquiritigenin, a flavonoid from licorice, blocks M2 macrophage polarization in colitis-associated tumorigenesis through downregulating PGE 2 and IL-6. Toxicology and Applied Pharmacology. 2014;**279**:311-321. DOI: 10.1016/j.taap.2014.07.001

[51] Sumiyoshi M, Taniguchi M, Baba K, Kimura Y. Antitumor and antimetastatic actions of xanthoangelol and 4-hydroxyderricin isolated from Angelica keiskei roots through the inhibited activation and differentiation of M2 macrophages. Phytomedicine. 2015;**22**(7-8):759-767. DOI: 10.1016/j.phymed.2015.05.005

[52] Wang YL, Tan X, Yang XL, Li XY, Bian K, Zhang DD. Total flavonoid from *Glycyrrhizae radix et rhizoma*

and its ingredient isoliquiritigenin regulation M2 phenotype polarization of macrophages. Zhongguo Zhong Yao Za Zhi. 2015;**40**(22):4475-4481

[53] Jang Y, Lee JK, Jeon YK, Kim CW. Exosome derived from epigallocatechin gallate treated breast cancer cells suppresses tumor growth by inhibiting tumor-associated macrophage infiltration and M2 polarization. BMC Cancer. 2013;**13**:421. DOI: 10.1186/1471-2407-13-421

[54] Park SA, Surh YJ. Modulation of tumor microenvironment by chemopreventive natural products. Annals of the New York Academy of Sciences. 2017;**1401**(1):65-74. DOI: 10.1111/nyas.13395

[55] Wang J, Man GCW, Chan TH, Kwong J, Wang CC. A prodrug of green tea polyphenol (−)-epigallocatechin-3-gallate (pro-EGCG) serves as a novel angiogenesis inhibitor in endometrial cancer. Cancer Letters. 2018;**412**:10-20. DOI: 10.1016/j.canlet.2017.09.054

[56] Choi HJ, Choi HJ, Chung TW, Ha KT. Luteolin inhibits recruitment of monocytes and migration of Lewis lung carcinoma cells by suppressing chemokine (CeC motif) ligand 2 expression in tumor-associated macrophage. Biochemical and Biophysical Research Communications. 2016;**470**(1):101-106. DOI: 10.1016/j.bbrc.2016.01.002

[57] Lee YJ, Lim T, Han MS, et al. Anticancer effect of luteolin is mediated by downregulation of TAM receptor tyrosine kinases,but not interleukin-8, in non-small cell lung cancer cells. Oncology Reports. 2017;**37**(2):1219-1226. DOI: 10.3892/or.2016.5336

[58] Fang B, Chen X, Wu M, Kong H, Chu G, Zhou Z, et al. Luteolin inhibits angiogenesis of the M2-like TAMs via the downregulation of hypoxia inducible factor-1a and the STAT3 signalling pathway under hypoxia. Molecular Medicine Reports. 2018;**18**(3):2914-2922. DOI: 10.3892/mmr.2018.9250

[59] Zhao X, Qu J, Liu X, Wang J, Ma X, Zhao X, et al. Baicalein suppress EMT of breast cancer by mediating tumor-associated macrophages polarization. American Journal of Cancer Research. 2018;**8**(8):1528-1540. eCollection 2018

[60] Zheng X, Meng WD, Xu YY, Cao JG, Qing FL. Synthesis and anticancer effect of chrysin derivatives. Bioorganic & Medicinal Chemistry Letters. 2003;**13**:881-884

[61] Sun S, Cui Y, Ren K, Quan M, Song Z, Zou H, et al. 8-bromo-7-methoxychrysin reversed M2 polarization of tumor-associated macrophages induced by liver cancer stem-like cells. Anti-Cancer Agents in Medicinal Chemistry. 2017;**17**(2):286-293

[62] Jiang YX, Chen Y, Yang Y, Chen XX, Zhang DD. Screening five Qi-Tonifying herbs on M2 phenotype macrophages. Evidence-based Complementary and Alternative Medicine. 2019;**2019**:9549315. DOI: 10.1155/2019/9549315

[63] Fujiwara Y, Saito Y, Shiota T, Cheng P, Ikeda T, Ohnishi K, et al. Natural compounds that regulate lymph node sinus macrophages: Inducing an anti-tumor effect by regulating macrophage activation. Journal of Clinical and Experimental Hematopathology. 2018;**58**(1):17-23. DOI: 10.3960/jslrt.17032

[64] Kang H, Zhang J, Wang B, Liu M, Zhao J, Yang M, et al. Puerarin inhibits M2 polarization and metastasis of tumor-associated macrophages from NSCLC xenograft model via inactivating MEK/ERK 1/2 pathway. International Journal of Oncology.

2017;**50**(2):545-554. DOI: 10.3892/ijo.2017.3841

[65] Ning Y, Feng W, Cao X, Ren K, Quan M, Chen A, et al. Genistein inhibits stemness of SKOV3 cells induced by macrophages co-cultured with ovarian cancer stem-like cells through IL-8/STAT3 axis. Journal of Experimental and Clinical Cancer Research. 2019;**38**:19. DOI: 10.1186/s13046-018-1010-1

[66] Xue N, Zhou Q, Ji M, Jin J, Lai F, Chen J, et al. Chlorogenic acid inhibits glioblastoma growth through repolarizating macrophage from M2 to M1 phenotype. Scientific Reports. 2017;**7**:39011. DOI: 10.1038/srep39011

[67] Oršolić N, Kunštić M, Kukolj M, Gračan R, Nemrava J. Oxidative stress, polarization of macrophages and tumour angiogenesis: Efficacy of caffeic acid. Chemico-Biological Interactions. 2016;**256**:111-124. DOI: 10.1016/j.cbi.2016.06.027

[68] Lee K, Ahn JH, Lee KT, Jang DS, Choi JH. Deoxyschizandrin, isolated from Schisandra berries, induces cell cycle arrest in ovarian cancer cells and inhibits the Protumoural activation of tumour-associated macrophages. Nutrients. 2018;**10**(1):91. DOI: 10.3390/nu10010091

[69] Sun L, Chen B, Jiang R, Li J, Wang B. Resveratrol inhibits lung cancer growth by suppressing M2-like polarization of tumor associated macrophages. Cellular Immunology. 2017;**311**:86-93. DOI: 10.1016/j.cellimm.2016.11.002

[70] Kimura Y, Sumiyoshi M. Resveratrol prevents tumor growth and metastasis by inhibiting Lymphangiogenesis and M2 macrophage activation and differentiation in tumor-associated macrophages. Nutrition and Cancer. 2016;**68**(4):667-678. DOI: 10.1080/01635581.2016.1158295

[71] Kimura Y, Sumiyoshi M, Baba K. Antitumor and antimetastatic activity of synthetic hydroxystilbenes through inhibition of lymphangiogenesis and M2 macrophage differentiation of tumor-associated macrophages. Anticancer Research. 2016;**36**:137-148

[72] Jeong SK, Yang K, Park YS, Choi YJ, Oh SJ, Lee CW, et al. Interferon gamma induced by resveratrol analog, HS-1793, reverses the properties of tumor associated macrophages. International Immunopharmacology. 2014;**22**(2):303-310. DOI: 10.1016/j.intimp.2014.07.004

[73] Mohammadi A, Blesso C, Barreto G, Banach M, Majeed M, Sahebkar M. Macrophage plasticity, polarization and function in response to curcumin, a diet-derived polyphenol, as an immunomodulatory agent. The Journal of Nutritional Biochemistry. 2019;**66**:1-16. DOI: 10.1016/j.jnutbio.2018.12.005

[74] Gao S, Zhou J, Liu N, Wang L, Gao Q, Wu Y, et al. Curcumin induces M2 macrophage polarization by secretion IL-4 and/or IL-13. Journal of Molecular and Cellular Cardiology. 2015;**85**:131-139. DOI: 10.1016/j.yjmcc.2015.04.025

[75] Tu SP, Jin H, Shi JD, Zhu LM, Suo Y, Lu G, et al. Curcumin induces the differentiation of myeloid-derived suppressor cells and inhibits their interaction with cancer cells and related tumor growth. Cancer Prevention Research (Philadelphia, Pa.). 2012;**5**(2):205-215. DOI: 10.1158/1940-6207.CAPR-11-0247

[76] Mukherjee S, Baidoo J, Fried A, Atwi D, Dolai S, Boockvar J, et al. Curcumin changes the polarity of tumor-associated microglia and eliminates glioblastoma. International Journal of Cancer. 2016;**139**(12):2838-2849. DOI: 10.1002/ijc.30398

[77] Mukherjee S, Hussaini R, White R, Atwi D, Fried A, Sampat S, et al.

TriCurin, a synergistic formulation of curcumin, resveratrol, and epicatechin gallate, repolarizes tumor-associated macrophages and triggers an immune response to cause suppression of HPV+ tumors. Cancer Immunology, Immunotherapy. 2018;**67**(5):761-774. DOI: 10.1007/s00262-018-2130-3

[78] Mukherjee S, Baidoo JNE, Sampat S, Mancuso A, David L, Cohen LS. Liposomal TriCurin, a synergistic combination of curcumin, epicatechin gallate and resveratrol, repolarizes tumor-associated microglia/macrophages, and eliminates glioblastoma (GBM) and GBM stem cells. Molecules. 2018;**23**;(1). pii: E201. DOI: 10.3390/molecules23010201

[79] Yao J, Du Z, Li Z, Zhang S, Lin Y, Li H, et al. 6-Gingerol as an arginase inhibitor prevents urethane-induced lung carcinogenesis by reprogramming tumor supporting M2 macrophages to M1 phenotype. Food & Function. 2018;**9**(9):4611-4620. DOI: 10.1039/c8fo01147h

[80] Wu Q, Chen GL, Li YJ, Chen Y, Lin FZ. Paeoniflorin inhibits macrophage-mediated lung cancer metastasis. Chinese Journal of Natural Medicines. 2015;**13**(12):925-932. DOI: 10.1016/S1875-5364(15)30098-4

[81] Shu G, Jiang S, Mu J, Yu H, Duan H, Deng X. Antitumor immunostimulatory activity of polysaccharides from *Panax japonicus* C. A. Mey: Roles of their effects on CD4 + T cells and tumor associated macrophages. International Journal of Biological Macromolecules. 2018;**111**:430-439. DOI: 10.1016/j.ijbiomac.2018.01.011

[82] Chen Y, Bi L, Luo H, Jiang Y, Chen F, Wang Y, et al. Water extract of ginseng and astragalus regulates macrophage polarization and synergistically enhances DDP's anticancer effect. Journal of Ethnopharmacology. 2019;**232**:11-20. DOI: 10.1016/j.jep.2018.12.003

[83] Li Q, Hao Z, Hong Y, He W, Zhao W. Reprogramming tumor associated macrophage phenotype by a polysaccharide from Ilex asprella for sarcoma immunotherapy. International Journal of Molecular Sciences. 2018;**19**(12). pii: E3816. DOI: 10.3390/ijms19123816

[84] Li Y, Qi W, Song X, Lv S, Zhang H, Yang Q. Huaier extract suppresses breast cancer via regulating tumor-associated macrophages. Scientific Reports. 2016;**6**:20049. DOI: 10.1038/srep20049

[85] Kimura Y, Sumiyoshi M. Antitumor and antimetastatic actions of dihydroxycoumarins (esculetin or fraxetin) through the inhibition of M2 macrophage differentiation in tumor-associated macrophages and/or G 1 arrest in tumor cells. European Journal of Pharmacology. 2015;**746**:115-125. DOI: 10.1016/j.ejphar.2014.10.048

[86] Pang L, Han S, Jiao Y, Jiang S, He X, Li P. Bu Fei decoction attenuates the tumor associated macrophage stimulated proliferation, migration, invasion and immunosuppression of non-small cell lung cancer, partially via IL-10 and PD-L1 regulation. International Journal of Oncology. 2017;**51**:25-38. DOI: 10.3892/ijo.2017.4014

[87] Iwanowycz S, Wang J, Hodge J, Wang Y, Yu F, Fan D. Emodin inhibits breast cancer growth by blocking the tumor-promoting feedforward loop between cancer cells and macrophages. Molecular Cancer Therapeutics. 2016;**15**(8):1931-1942. DOI: 10.1158/1535-7163.MCT-15-0987

[88] Jia X, Yu F, Wang J, Iwanowycz S, Saaoud F, Wang Y, et al. Emodin suppresses pulmonary metastasis of breast cancer cells accompanied with

decreased macrophage recruitment and M2 polarization in the lungs. Breast Cancer Research and Treatment. 2014;**148**(2):291-302. DOI: 10.1007/s10549-014-3164-7

[89] Iwanowycz S, Wang J, Altomare D, Hui Y, Fan D. Emodin bidirectionally modulates macrophage polarization and epigenetically regulates macrophage memory. The Journal of Biological Chemistry. 2016;**291**(22):11491-11503. DOI: 10.1074/jbc.M115.702092

[90] Morris J, Gonzales CB, De La Chapa JJ, Cabang B, Fountzilas C, Patel M, et al. The highly pure neem leaf extract, SCNE, inhibits tumorigenesis in oral squamous cell carcinoma via disruption of pro-tumor inflammatory cytokines and cell signaling. Frontiers in Oncology. 2019;**9**:890. DOI: 10.3389/fonc.2019.00890

[91] Yu X, Xu M, Li N, Li Z, Li H, Shao S, et al. β-Elemene inhibits tumor-promoting effect of M2 macrophages in lung cancer. Biochemical and Biophysical Research Communications. 2017;**490**(2):514-520. DOI: 10.1016/j.bbrc.2017.06.071

[92] Tsuboki J, Fujiwara Y, Horlad H, Shiraishi D, Nohara T, Tayama S, et al. Onionin a inhibits ovarian cancer progression by suppressing cancer cell proliferation and the protumour function of macrophages. Scientific Reports. 2016;**6**:29588. DOI: 10.1038/srep29588

[93] Sinha P, Ostrand-Rosenberg S. Myeloid-derived suppressor cell function is reduced by Withaferin a, a potent and abundant component of *Withania somnifera* root extract. Cancer Immunology, Immunotherapy. 2013;**62**(11):1663-1673. DOI: 10.1007/s00262-013-1470-2

[94] Woo SM, Choi YK, Cho SG, Park S, Ko SG. A new herbal formula, KSG-002, suppresses breast cancer growth and metastasis by targeting NF-κ B-Dependent TNF α production in macrophages. Evidence-Based Complementary and Alternative Medicine. 2013;**2013**:728258. DOI: 10.1155/2013/728258

[95] Choi YK, Cho SG, Woo SM, Yun YJ, Park S, Shin YC, et al. Herbal extract SH003 suppresses tumor growth and metastasis of MDA-MB-231 breast cancer cells by inhibiting STAT3-IL-6 signaling. Mediators of Inflammation. 2014;**2014**:492173. DOI: 10.1155/2014/492173

[96] Wu J, Zhang XX, Wang M, Wang HX, Wang YH, Li C, et al. The effect of Jianpi Yangzheng Xiaozheng decoction and its components on gastric cancer. Journal of Ethnopharmacology. 2019;**10**(235):56-64. DOI: 10.1016/j.jep.2019.02.003

[97] Wang N, Zheng Y, Gu J, Cai Y, Wang S, Zhan F, et al. Network-pharmacology-based validation of TAMS/CXCL-1 as key mediator of XIAOPI formula preventing breast cancer development and metastasis. Scientific Reports. 2017;7(1):14513. DOI: 10.1038/s41598-017-15030-3

Chapter 3

Functional Biomaterials Modulate Macrophage in the Tumour Micro-environment

Tsung-Meng Wu, Kuang-Teng Wang, Hisang-Lin Tsai, Fan-Hua Nan and Yu-Sheng Wu

Abstract

The inflammation response requires the cooperation of macrophages with immune cell function and active factors, such as cytokines and chemokines. Through this response, these factors are involved in the immune response to affect physiological activities. Macrophages can be categorized into two types: 'M1' and 'M2'. M1 macrophages destroy the pathogen through phagocytosis activation, ROS production, and antigen-presenting, among other functions. M2 macrophages release cellular factors for tissue recovery, growth, and angiogenesis. Studies have determined that tumour tissue presents with numerous macrophages, termed tumour-associated macrophages. Tumour cells and peripheral stromal cells stimulate the tumour associated with macrophages (M2) to produce factors that regulate angiogenesis. Modulating the balance of the M1 and M2 function has already gained interest as a potentially valuable immune disease therapy. However, applications of the immunotherapy in clinical treatments are still not clear with regard to the cellular working mechanism. Therefore, we summarized the functions of common biomaterials involved in the modulation of the macrophage.

Keywords: macrophage, polarization, tumour micro-environment, biomaterials, cytokines

1. Introduction

Inflammation has been demonstrated to be a critical factor in the induction of immune disease. Immunotherapy is a novel therapeutic approach for anti-inflammation, which could help avoid drug resistance. However, findings have indicated that the balance of inflammation and anti-inflammation is crucial. Cellular ROS are produced by stress to clear pathogen infections [1]. The inflammatory response involves macrophages, dendritic cells, and M cells, which are crucial protectors. These cells present partial antigens to enhance the T-cell activation and cytokine production, which modulate the host micro-environment. Cytokines are produced and released as signals to regulate the immune cell function.

Immunotherapy was developed as an approach to rectify the imbalanced inflammation. Immunotherapy was hypothesized as a possible alternative therapy applied in the early phase of clinical therapy and immunomodulation in the early stages of immune disease. The common immunotherapy employs natural functional materials including triterpenoids and polysaccharides. Studies have demonstrated that functional polysaccharides can promote macrophage differentiation into M1 or M2, and the ratio modulates the host micro-environment through cytokine secretion.

Polysaccharides such as beta-glucan are considered to be biological response modifiers (BRMs) that activate macrophages and modulate the inflammation response. Findings have indicated that beta-glucan combines with receptors expressed on the macrophage cell surface, such as Toll-like receptor. Once combined, alveolar macrophages, Kupffer cells, Langerhans cells, mesangial cells, and microglial are activated through toll-like receptor 4–mediated signalling pathways to modulate the immune response.

2. Macrophage activation

Macrophages are present in almost all tissues and coordinate developmental, metabolic, and immunological functions, thereby contributing to the maintenance of homeostasis. Macrophages have a complex role in tissues and act on lipopolysaccharide (LPS), interferon-γ (IFN-γ), and interleukin (IL)-4 to polarize the M0 into M1. Macrophages are activated by exposure to various stimuli. The stimuli that act on macrophages are categorized into danger, homeostatic, metabolic, and modulatory signals. Danger signals include pathogen-associated molecular patterns, such as LPS. Tissue macrophage exposure to danger signals results in an inflammatory response. Findings have indicated that tumour environments contain numerous transmitters, such as M-CSF, IL-6, IL-10, TGF-β, and COX-2, which induce tumour megakaryocytes to differentiate into M2 macrophages, which, in addition to having poorer antigen-presenting and cytotoxic abilities, also secrete factors that inhibit immune cells, resulting in an enhanced immune inhibitory effect of the tumour environment as shown in **Figure 1**. We investigated the modulation of M1 and M2 in the tumour environment by using immunomodulators to delay or inhibit the tumour to identify alternative approaches to reduce the side effect of tumour chemotherapy. Inflammation is a crucial adaptive response for animals, and the mechanism involves a complex interaction of molecular mediators. The functions of immune cells in a micro-environment are mediated by responses that occur at all levels of biological organization [2]. This process involves cooperation among cells and mediators, and the classical immune response varies based on a wide range of factors, including the stage of the inflammation process, the tissue or organ involved, and whether the inflammation is acute and resolving or chronic and nonresolving [3]. The inflammation process involves vascular permeabilization, active migration of blood cells, and passage of plasma constituents into injurious tissue [4]. Studies have demonstrated that the infiltration of immune cells during the inflammation process plays a crucial role in atherosclerosis [5]. Blood leukocytes are mediators of host defences and inflammation localized in the earliest lesions of atherosclerosis in experimental animals. The study of inflammation in atherosclerosis provided new insights into the mechanisms underlying the recruitment of leukocytes [6]. Recently, studies have indicated that inflammation plays a role in Alzheimer disease (AD) [7]. Inflammatory components involved in AD neuroinflammation include brain cells (such as microglia and astrocytes), the complement system, and cytokines and chemokines [8]. Regarding cancer development [9], proinflammatory cytokines, including chemokines; matrix metallopeptidase (MMP)-9; vascular endothelial growth factor (VEGF); and IL-1α, IL-1β, IL-6, IL-8, and IL-18, are primarily regulated by the transcription factor nuclear factor (NF)-kB, which is active in most tumours and is induced by carcinogens [10]. Cutaneous wound repair

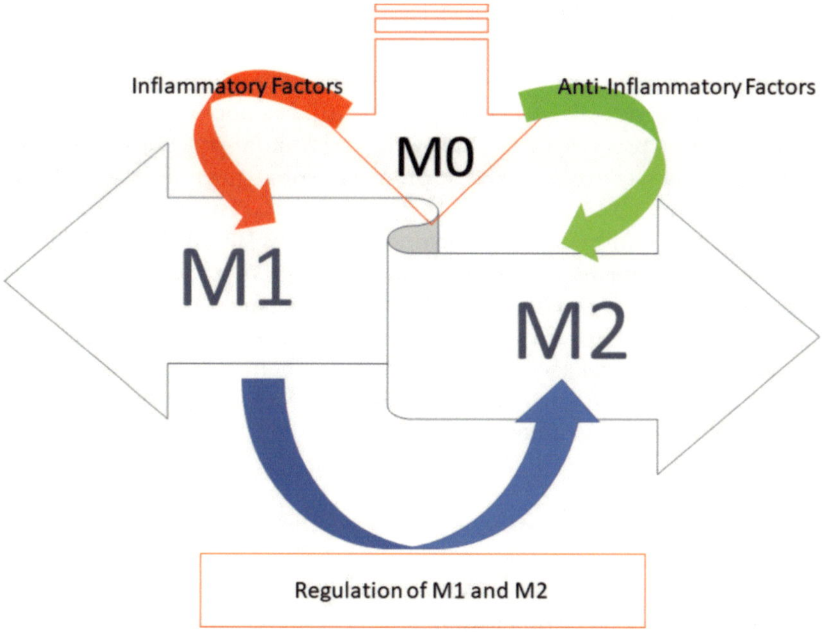

Figure 1.
Macrophages can be categorized into two types: 'M1' and 'M2'. M1 macrophages destroy the pathogen through phagocytosis activation, ROS production, and antigen-presenting, among other functions. M2 macrophages release cellular factors for tissue recovery, growth, and angiogenesis. We thought that the regulation of macrophage is beneficial to reduce the auto-immune disease.

is a tightly regulated and dynamic process involving blood clotting, inflammation, new tissue formation, and tissue remodeling [11]. Thrombin is the protease involved in blood coagulation. Thrombin deregulation can lead to haemostatic abnormalities, which range from subtle subclinical to life-threatening coagulopathies (i.e., during septicaemia) [12]. Inflammation and blood coagulation is part of the innate host protection mechanism against vascular injury, infection, or other wounds. Cells of the innate immune system, endothelial cells, and platelets are actively involved in acute and chronic inflammation; they release proinflammatory mediators and recruit leukocytes [13]. The protease-activated receptor (PAR) family serves as sensor of serine proteinases in the blood clotting system in the target cells involved in inflammation. Activation of PAR-1 by thrombin and of PAR-2 by factor leads to a rapid expression and exposure of both adhesive proteins that mediate an acute inflammatory reaction and of the tissue factor that initiates the blood coagulation cascade on the membrane of endothelial cells [14]. In this process, cooperation among cells and mediators occurs, and a wide range of factors are involved in the classical immune response: (1) the stage of the inflammation process; (2) the tissue or organ involved; and (3) whether the inflammation is acute and resolving or chronic and nonresolving [15]. The inflammation process involves vascular permeability, active migration of blood cells, and the passage of plasma constituents into injurious tissue [4]. Studies on the infiltration of immune cells have demonstrated that the inflammation process plays a crucial role in atherosclerosiss [5]. Blood leukocytes, mediators of host defences and inflammation, localize in the earliest lesions of atherosclerosis in experimental animals. The study of inflammation in atherosclerosis has provided numerous new insights into the mechanisms underlying the recruitment of leukocytes [6]. Studies have reported that inflammation is involved in Alzheimer's disease (AD) [7]. Inflammatory components involved in AD neuroinflammation include brain cells (such as microglia and astrocytes), the complement system, and cytokines and chemokines [8]. Regarding cancer development, proinflammatory cytokines, including chemokines, MMP-9, VEGF, and IL-1α, IL-1β, IL-6, IL-8, and IL-18, are primarily regulated by the transcription factor NF-kB, which is active in most tumours and is induced by carcinogens [9, 10]. Macrophages play a crucial role in inflammation process, tumour growth, and tumour progression by induced angiogenesis. Studies have reported that promotion of angiogenesis with the production of proangiogenic factors, such as TGFβ, VEGF, PDGF, members of the fibroblast growth factors family, and angiogenic chemokines [16], and the development of breast cancer and several other human tumours was correlated with macrophage infiltration [17]. VEGF-C production by tumour-associated macrophages (TAMs) was reportedly involved in peritumoral lymphangiogenesis and the subsequent dissemination of cancer cells with formation of lymphatic metastasis [18]; moreover, macrophage colony-stimulating factor (M-CSF) and VEGF actively recruit circulating blood monocytes at the tumour site [19].

3. Polysaccharide function on the immunomodulation

Evidence has indicated that acetyl-xylogalactan extracted from *Sarcodia suieae* induced macrophage polarization through the IL-1β, TNF, and Malt-1 expression [20]. Nakanishi et al determined that celecoxib can alter the immune inhibitory effects of the tumour micro-environment by promoting the transformation of TAMs into M1 macrophages, leading to inhibited tumour growth [21]. In 1968, Ikekawa et al. reported that the fruiting body extracts from *Lentinus edodes*, *Trametes versicolor*, *Ganoderma tsugae*, *Flammulina velutiper,* and *Tricholoma matsutake* demonstrated significant antitumour activities in transplanted Sarcoma 180 tumour cells [22, 23]. Studies have reported that *Antrodia camphorata*-derived beta-glucan demonstrated inhibitory effects on tumour growth in Sarcoma 37, Sarcoma 180, Erlich ascites sarcoma, Yoshida sarcoma, and LLC1 transplanted tumour [24]. Daily intake of *A. comphorata*-derived beta-glucan for 18 consecutive days was demonstrated to slow tumour growth and reduce the rate of metastasis [25]. Cytotoxic T-cell activity and tumour occurrence rate were investigated and the results revealed that daily oral intake of *Grifola frondosa*-derived beta-glucan or Lentinan can enhance cytotoxic T-cell activity and reduce tumour occurrence rate [26]. Furthermore, the addition of conditioned medium with tumour cells into the progenitors of dendritic cells was determined to further inhibit the maturation of dendritic cells and lower the antigen-presenting capability of the dendritic cells [27]. Studies have reported that tumour cells secrete M-CSF, thereby inhibiting dendritic and T-cell differentiation and antitumour ability [27–30]. In the tumour environment, the amounts of M1 and M2 macrophages are not equal [31]. Tumour environments are known to contain a large number of transmitters such as M-CSF, IL-6, IL-10, TGF-β, and COX-2, which induce tumour megakaryocytes to differentiate into M2 macrophages, which have poorer antigen-presenting and cytotoxic abilities and secrete factors that inhibit immune cells, resulting in an enhanced immune inhibitory effect of the tumour environment [16, 32–41]. M2 macrophages in tumour-bearing mice enhance tumour growth and immune inhibitory effects. M2 macrophages also secrete cytokines, such as IL-10 and TGF-β, in high quantities, which attract noncytotoxic Treg-cells and Type 2 helper T cells to congregate in tumour tissues, which in turn inhibit the differentiation and normal functions of T cells, including their cytotoxic ability, which further leads to T-cell apoptosis [38, 40, 42–44]. The Th1 and Th2 polarization is built on cytokine patterns, which begin when the antigen-presenting cells interact with the naive T cells and polarize into type 1 and type 2 cells in response to the type of antigen encountered [45]. Th1 and Th2 cells secrete different cytokines.

Th1 cells are dependent on IL-2, IFN-γ, and TNF, which are involved in cell-mediated immunity against pathogens. Th2 cells are mostly dependent on IL-4 and IL-5, which stimulate the production of IgE antibodies and eosinophil responses, resulting in allergic diseases [46, 47]. An imbalanced Th1/Th2 immune response is linked to certain hypersensitivity disorders such as allergy, asthma, and hay fever [48]; therefore, studies have suggested that using BRM to restore the balance between Th1 and Th2 immune response could be a treatment option for the IgE-dependent hypersensitivity [49]. *Ganoderma lucidum* is a medicinal mushroom, which has been widely used for hundreds of years as a folk medicine in oriental countries such as China and Japan for its immunomodulating and antitumour effects. Numerous biological available substances with immunity enhancement effects, in particular polysaccharides, have been isolated from the extract of *G. lucidum* [50].

Antimicrobial peptides are effective components of innate immunity that are widely present in the biological system. Hepcidin is a 25-amino acid antibiotic peptide synthesized in the liver, which is reportedly responsible for regulating iron balance and recycling in humans and mice. Studies on 0–100 μg/mL concentrations of hepcidin incubated with HT1080, Hep-G2, and HeLa for 24 h revealed higher growth inhibition ratios after treatment with 70 μg/mL hepcidin in HT1080 cells. Hepcidin was very effective at inhibiting the growth of fibrosarcoma cells [51, 52]. Studies on tachyplesin, an antimicrobial peptide present in leukocytes of the horseshoe crab (*Tachypleus tridentatus*), demonstrated that tachyplesin was able to inhibit the growth of TSU tumour cells on the chorioallantoic membrane of chicken embryos and B16 tumour cells in syngeneic mice. Tachyplesin also blocked the proliferation of both tumour and endothelial cells in culture in a dose-dependent manner, whereas proliferation was relatively unaffected in nontumorigenic cell lines Cos-7 and NIH-3T3 [53]. D-K4R2L9 is a peptide with 15 amino acid residues, comprising of Leu, Lys, and Arg residues, which binds to and lyses B16-F10 mouse melanoma cells in culture at concentrations that do not harm normal 3T3 fibroblasts or erythrocytes, thereby preventing intravenous-injected D122 lung carcinoma cells from forming lung tumours in mice [54, 55]. Another antimicrobial peptide, bovine lactoferricin (LfcinB), is a 25-amino acid, highly basic peptide with a disulphide bridge between two cysteines; thus, LfcinB is a cyclic twisted antiparallel β-sheet solution structure. The effects of LfcinB on neuroblastoma growth were investigated in vivo, which revealed that SH-SY-5Y xenografts in nude rats were significantly inhibited after injections of 1.0 or 2.0 mg of LfcinB compared with untreated controls [56]. Related research has demonstrated that antimicrobial peptides can activate specific innate immune responses and lead to immunomodulatory effects in the host when there is a risk of damage. Furthermore, the antimicrobial peptides are proposed to modulate the host's immune system through inflammatory responses and stimulate the beneficial aspects of inflammation, including inhibition of tumour growth.

4. Conclusion

Immunotherapy is being developed and presents certain advantages of alternative medicine because immunomodulation factors, such as mushroom beta-glucan, antimicrobial peptides, and triterpenoid, represent a novel therapeutic approach for cancer therapy and may provide an alternative to deal with the problem of drug resistance. However, exploring current insights into tumour biology and tumour micro-environment is complex and involves chemistry, biology, instrumentation, and formulation science. Therefore discovering a novel, more effective tumour-targeting treatment is difficult. Immunotherapy is hypothesized to be an alternative therapy that could be applied in the early phase of clinical tumour therapy.

Competing interests

The authors declare no competing interests.

Author details

Tsung-Meng Wu[1], Kuang-Teng Wang[1], Hisang-Lin Tsai[2], Fan-Hua Nan[3] and Yu-Sheng Wu[1]*

1 Department of Aquaculture, National Pingtung University of Science and Technology, Pingtung, Taiwan

2 Surgery Medical College of Kaohsiung Medical University, Kaohsiung, Taiwan

3 Department of Aquaculture, National Taiwan Ocean University, Keelung, Taiwan

*Address all correspondence to: wuys0313@mail.npust.edu.tw

References

[1] Rossi FR, Krapp AR, Bisaro F, Maiale SJ, Pieckenstain FL, Carrillo NJTPJ. Reactive oxygen species generated in chloroplasts contribute to tobacco leaf infection by the necrotrophic fungus *Botrytis cinerea*. The Plant Journal. 2017;**92**:761-773

[2] Allavena P, Sica A, Solinas G, Porta C, Mantovani A. The inflammatory micro-environment in tumor progression: The role of tumor-associated macrophages. Critical Reviews in Oncology/Hematology. 2008;**66**:1-9

[3] Punchard NA, Whelan CJ, Adcock I. The journal of inflammation. Journal of Inflammation (London). 2004;**1**:1

[4] Maslinska D, Gajewski M. Some aspects of the inflammatory process. Folia Neuropathologica. 1998;**36**:199-204

[5] Sbarsi I, Falcone C, Boiocchi C, Campo I, Zorzetto M, De Silvestri A, et al. Inflammation and atherosclerosis: The role of tnf and tnf receptors polymorphisms in coronary artery disease. International Journal of Immunopathology and Pharmacology. 2007;**20**:145-154

[6] Libby P, Ridker PM, Maseri A. Inflammation and atherosclerosis. Circulation. 2002;**105**:1135-1143

[7] Schott JM, Revesz T. Inflammation in alzheimer's disease: Insights from immunotherapy. Brain. 2013;**136**:2654-2656

[8] Rubio-Perez JM, Morillas-Ruiz JM. A review: Inflammatory process in Alzheimer's disease, role of cytokines. Scientific World Journal. 2012;**2012**:1-15

[9] Gregory CD. Inflammation and cancer revisited: An hypothesis on the oncogenic potential of the apoptotic tumor cell. Autoimmunity. 2013;**46**:312-316

[10] Aggarwal BB, Shishodia S, Sandur SK, Pandey MK, Sethi G. Inflammation and cancer: How hot is the link? Biochemical Pharmacology. 2006;**72**:1605-1621

[11] Muller AK, Meyer M, Werner S. The roles of receptor tyrosine kinases and their ligands in the wound repair process. Seminars in Cell & Developmental Biology. 2012;**23**:963-970

[12] Danckwardt S, Hentze MW, Kulozik AE. Pathologies at the nexus of blood coagulation and inflammation: Thrombin in hemostasis, cancer, and beyond. Journal of Molecular Medicine. 2013;**91**:1257-1271

[13] Strukova S. Blood coagulation-dependent inflammation. Coagulation-dependent inflammation and inflammation-dependent thrombosis. Frontiers in Bioscience (Elite Edition). 2006;**11**:59-80

[14] Dugina TN, Kiseleva EV, Chistov IV, Umarova BA, Strukova SM. Receptors of the par-family as a link between blood coagulation and inflammation. Biochemistry (Moscow). 2002;**67**:65-74

[15] Zaenker KS. Journal of inflammation & allergy—Drug targets. Inflammation & Allergy Drug Targets. 2009;**8**:1

[16] Mantovani A, Sozzani S, Locati M, Allavena P, Sica A. Macrophage polarization: Tumor-associated macrophages as a paradigm for polarized m2 mononuclear phagocytes. Trends in Immunology. 2002;**23**:549-555

[17] Leek RD, Lewis CE, Whitehouse R, Greenall M, Clarke J, Harris AL. Association of macrophage infiltration with angiogenesis and prognosis in invasive breast carcinoma. Cancer Research. 1996;**56**:4625-4629

[18] Schoppmann SF, Birner P, Stockl J, Kalt R, Ullrich R, Caucig C, et al. Tumor-associated macrophages express lymphatic endothelial growth factors and are related to peritumoral lymphangiogenesis. The American Journal of Pathology. 2002;**161**:947-956

[19] Sica A, Schioppa T, Mantovani A, Allavena P. Tumour-associated macrophages are a distinct m2 polarised population promoting tumour progression: Potential targets of anti-cancer therapy. European Journal of Cancer. 2006;**42**:717-727

[20] Wu T-M, Nan F-H, Chen K-C, Wu Y-S. *Sarcodia suieae* acetyl-xylogalactan regulate raw 264.7 macrophage nf-kappa b activation and il-1 beta cytokine production in macrophage polarization. Scientific Reports. 2019;**9**:1-10

[21] Nakanishi Y, Nakatsuji M, Seno H, Ishizu S, Akitake-Kawano R, Kanda K, et al. Cox-2 inhibition alters the phenotype of tumor-associated macrophages from m2 to m1 in apcmin/+ mouse polyps. Carcinogenesis. 2011;**32**:1333-1339

[22] lkekawa T. Enokitake, flammulina velutipes: Host-mediated antitumor polysaccharides. Food Reviews International. 1995;**11**:203-206

[23] Ikekawa T, Uehara N, Maeda Y, Nakanishi M, Fukuoka F. Antitumor activity of aqueous extracts of edible mushrooms. Cancer Research. 1969;**29**:734-735

[24] Wasser SP. Medicinal mushrooms as a source of antitumor and immunomodulating polysaccharides. Applied Microbiology and Biotechnology. 2002;**60**:258-274

[25] Inomata T, Goodman GB, Fryer CJ, Chaplin DJ, Palcic B, Lam GK, et al. Immune reaction induced by x-rays and pions and its stimulation by schizophyllan (spg). The British Journal of Cancer. 1996;**27**:122-125

[26] Nanba H, Kubo K. Effect of maitake d-fraction on cancer prevention. Cancer. 1997;**833**:204-207

[27] Menetrier-Caux C, Montmain G, Dieu MC, Bain C, Favrot MC, Caux C, et al. Inhibition of the differentiation of dendritic cells from cd34$^+$ progenitors by tumor cells: Role of interleukin-6 and macrophage colony-stimulating factor. Blood. 1998;**92**:4778-4791

[28] Lin EY, Gouon-Evans V, Nquyen AV, Pollard JW. The macrophage growth factor csf-1 in mammary gland development and tumor progression. Journal of Mammary Gland Biology and Neoplasia. 2002;**7**:147-162

[29] Lewis CE, Pollard JW. Distinct role of macrophages in different tumor microenvironment. Cancer Research. 2006;**66**:605-612

[30] Joyce JA, Pollard JW. Microenvironmental regulation of metastasis. Nature Reviews Cancer. 2009;**9**:239-252

[31] Ch'ng ES, Jaafar H, Tuan Sharif SE. Breast tumor angiogenesis and tumor-associated macrophages: Histopathologist's perspective. Pathology Research International. 2011;**2011**:1-13

[32] Elgert KD, Alleva DG, Mullins DW. Tumor-induced immune dysfunction: The macrophage connection. Journal of Leukocyte Biology. 1998;**64**:275-290

[33] Sunderkötter C, Goebeler M, Schulze-Osthoff K, Bhardwaj R, Sorg C. Macrophage-derived angiogenesis factors. Pharmacology & Therapeutics. 1991;**51**:195-216

[34] Giraudo E, Inoue M, Hanahan D. An amino-bisphosphonate targets mmp-9-expressing macrophages

and angiogenesis to impair cervical carcinogenesis. The Journal of Clinical Investigation. 2004;**114**:623-633

[35] Ben-Baruch A. Inflammation-associated immune suppression in cancer: The roles played by cytokines, chemokines and additional mediators. Seminars in Cancer Biology. 2006;**16**:38-52

[36] Mitsuhashi M, Liu J, Cao S, Shi X, Ma X. Regulation of interleukin-12 gene expression and its anti-tumor activities by prostaglandin e2 derived from mammary carcinomas. Journal of Leukocyte Biology. 2004;**76**:322-332

[37] Flavell RA, Sanjabi S, Wrzesinski SH, Licona-Limón P. The polarization of immune cells in the tumour environment by tgfbeta. Nature Reviews Immunology. 2010;**10**:554-567

[38] Martinez FO, Helming L, Gordon S. Alternative activation of macrophages: An immunologic functional perspective. Annual Review of Immunology. 2009;**27**:451-483

[39] Bingle L, Brown NJ, Lewis CE. The role of tumour-associated macrophages in tumour progression: Implications for new anticancer therapies. The Journal of Pathology. 2002;**196**:254-265

[40] Siemann DW. Tumor microenvironment. Wiley Online Library. 2011

[41] Pollard JW. Tumour-educated macrophages promote tumour progression and metastasis. Nature Reviews Cancer. 2004;**4**:71-78

[42] Standiford TJ, Kuick R, Bhan U, Chen J, Newstead M, Keshamouni VG. Tgf-β-induced irak-m expression in tumor-associated macrophages regulates lung tumor growth. Oncogene. 2011;**30**:2475-2484

[43] Murphy K, Travers P, Walport M. Immunobiology. 7th ed. USA: Graland Science; 2008

[44] Solinas G, Germano G, Mantovani A, Allavena P. Tumor-associated macrophages (tam) as major players of the cancer-related inflammation. Journal of Leukocyte Biology. 2009;**86**:1065-1073

[45] Mosmann TR, Cherwinski H, Bond MW, Giedlin MA, Coffman RL. Two types of murine helper t cell clone. I. Definition according to profiles of lymphokine activities and secreted proteins. Journal of Immunology. 2005;**175**:5-14

[46] Mosmann TR, Coffman RL. Th1 and th2 cells: Different patterns of lymphokine secretion lead to different functional properties. Annual Review of Immunology. 1989;**7**:145-173

[47] Maggi E. The th1/th2 paradigm in allergy. Immunotechnology. 1998;**3**:233-244

[48] Kidd P. Th1/th2 balance: The hypothesis, its limitations, and implications for health and disease. Alternative Medicine Review. 2003;**8**:223-246

[49] Stern ME, Siemasko KF, Niederkorn JY. The th1/th2 paradigm in ocular allergy. Current Opinion in Allergy and Clinical Immunology. 2005;**5**:446-450

[50] Lin ZB, Zhang HN. Anti-tumor and immunoregulatory activities of ganoderma lucidum and its possible mechanisms. Acta Pharmacologica Sinica. 2004;**25**:1387-1395

[51] Shi JS, Camus AC. Hepcidins in amphibians and fishes: Antimicrobial peptides or iron-regulatory hormones? Developmental and Comparative Immunology. 2006;**30**:746-755

[52] Chen JY, Lin WJ, Lin TL. A fish antimicrobial peptide, tilapia hepcidin th2-3, shows potent antitumor activity against human fibrosarcoma cells. Peptides. 2009;**30**:1636-1642

[53] Chen YX, Xu XM, Hong SG, Chen JG, Liu NF, Underhill CB, et al. Rgd-tachyples in inhibits tumor growth. Cancer Research. 2001;**61**:2434-2438

[54] Hoskin DW, Ramamoorthy A. Studies on anticancer activities of antimicrobial peptides. Biochimica et Biophysica Acta: Biomembranes. 2008;**1778**:357-375

[55] Papo N, Shahar M, Eisenbach L, Shai Y. A novel lytic peptide composed of dl-amino acids selectively kills cancer cells in culture and in mice. The Journal of Biological Chemistry. 2003;**278**:21018-21023

[56] Eliassen LT, Berge G, Leknessund A, Wikman M, Lindin I, Lokke C, et al. The antimicrobial peptide, lactoferricin b, is cytotoxic to neuroblastoma cells in vitro and inhibits xenograft growth in vivo. International Journal of Cancer. 2006;**119**:493-500

Section 2

Macrophage and Infection Control

Chapter 4

New Tools for Studying Macrophage Polarization: Application to Bacterial Infections

Soraya Mezouar and Jean-Louis Mege

Abstract

Macrophages are tissue immune cells involved in homeostasis and are considered as the first line of defense during bacterial infections. They are resident cells but may be recruited during inflammation and/or infection. Hence, their study is necessary not only to decipher innate immune mechanisms involved in bacterial infections but also to follow infected patients. Among the numerous functions of macrophages, their polarization into microbicidal or permissive cells has been an interesting concept to describe their responses to bacterial aggression. Numerous *in vitro* studies, including ours, have shown the ability of bacteria to induce different patterns of macrophage polarization. However, the studies of patients during infections have produced less convincing results. We propose in this review to take stock of the tools for studying the polarization of macrophages and to show their limits. We make recommendations for using macrophage polarization as a biomarker for measuring severity and response to treatment in bacterial infectious diseases.

Keywords: macrophage, infection, bacterium, polarization, infectious diseases

1. Introduction

Elie Metchnikoff used for the first time the term "macrophage" to describe highly mobile cells able to phagocyte bacteria, which earned him the Nobel Prize in 1908 [1]. During several decades, it was admitted that macrophages are issued from circulating monocytes in homeostatic conditions or after their migration to the tissues following chemokine gradients. More recently, the use of new tools such as genetic fate mapping techniques has shown that most of resident macrophages are of embryonic origin and monocytes contribute to their renewal when homeostasis is impaired [2]. In addition to their role in regulation of tissue development and homeostasis, macrophages actively participate to innate immune defense through the recognition of viruses, bacteria, parasites or fungi [3].

In contrast to lymphoid cells, macrophages are neither antigen specific nor clonally restricted and express a large panel of membrane molecules. The activation ways of macrophages during infection rely on the interaction of pathogen-associated molecular pattern (PAMPs) with pathogen recognition receptors (PRRs) such as toll-like receptors (TLRs), scavenger receptors, C type lectin receptors, or complement receptors [4]. The interaction between PAMPs and PRRs leads to the activation of macrophages including production and secretion of cytokines, chemokines and lipid mediators, and promotes the uptake of microorganisms and their destruction [5]. Hence, macrophages are at the center of anti-infectious immune response, which includes pathogen recognition, macrophage activation and pathogen elimination [6, 7].

The polarization state of macrophage is characterized by their activation by pathogen ligands and inflammatory molecules. As previously described for T cell subsets with Th1 and Th2 functional dichotomy, M1 and M2 polarization may correspond to downstream effects of T cell polarization [8]. Numerous approaches have been performed to investigate macrophage responses during infection. Among them, the concept of polarization profile has represented a powerful strategy to investigate macrophage activation states during infection [9]. Here, we investigate tools available to study macrophages in a critical point of view and we propose them to assess prognosis and therapeutic response in bacterial infectious diseases.

2. Concept of macrophage polarization

The term of "polarization" corresponds to functional states exhibiting a binary distribution. It was used for the first time in 1986 by Mosmann et al. to characterize two murine T helper lymphocyte sub-populations, i.e., Th1 and Th2 according to their respective stimuli, interferon (IFN)-γ and interleukin (IL)-4 [10]. The concept of macrophage polarization was deduced from the Th1 and Th2 polarization and accounted for the diversity of macrophage activation [8]. Hence, Stein et al. showed that IL-4 stimulated the expression by murine macrophages of the mannose type 1 receptor (MRC1, CD206) associated to enhanced particle uptake and decreased release of tumor necrosis factor (TNF), a potent inflammatory cytokine; these characteristics may be considered as a model of M2 signature [11]. Later, Mills et al. confirmed that Th1 or Th2 lymphocytes led to the polarization of macrophages into M1 (inflammatory) and M2 (immunoregulatory) profiles [8]. Another nomenclature coexisted with M1/M2 polarization: M1 macrophages were also called classically activated macrophages while M2 macrophages exhibited an alternative type of activation [12]. Few authors use now these two terms and the heterogeneity of M2 macrophages do not fit with the category of alternative activation, explaining why we will use only M1 and M2 terms.

As depicted in **Figure 1**, M1 and M2 profiles are induced by specific ligands. M1 profile is elicited by inflammatory cytokines (TNF or IFN-γ), bacterial components such as lipopolysaccharide (LPS) or growth factors including granulocyte-macrophage colony-stimulating factor (GM-CSF). In contrast, Th2 cytokines (IL-4, IL-10 and IL-13) lead to M2 polarization in the same way as IL-33, transforming growth factor (TGF)-β or macrophage colony-stimulating factor (M-CSF), the master growth factor of myeloid lineage. According the way of stimulation, macrophages express several different markers, secrete different mediators and exercise specific functions (**Figure 1**) [13–16]. It is important to note that M2 macrophages are more heterogeneous than M1 macrophages and have been divided into three distinct profiles including M2a, M2b and M2c according their functions as "alternative and repairers" (M2a) or anti-inflammatory regulators (M2b and M2c) [17, 18]. M2a, M2b and M2c macrophages are activated by IL-14 and IL-13, immunes complexes associated with TLRs or glucocorticoids, IL-10 and TGF-β, respectively [13]. In contrast to a general point of view, using mass spectrometry we found that IFN-γ-stimulated macrophages exhibit a proteomic profile distinct from LPS-stimulated macrophages or LPS/IFN-γ-stimulated macrophages even if they are all included in M1 category [19]. The appearance of numerous discrepancies with the concept of M1/M2 dichotomy led scientists working on macrophage polarization to propose a

	M1 polarization	M2 polarization
Stimulation	• Proinflammatory cytokine TNF • Lipopolysaccharide *via* TLR-4 • IFN-γ • GM-CSF	• Th2 cytokines: IL-4, IL-10 and IL-13 • Glucocorticoids • IL-33 • TGF-β • G-CSF
Markers and cytokines production	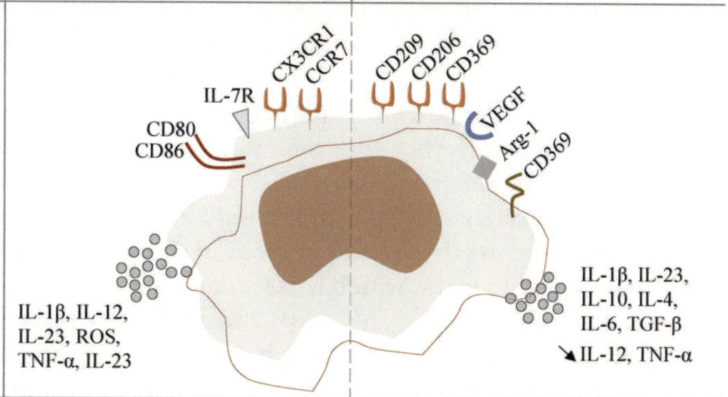	
Functions	• Progression of inflammation • Th1 response • Antimicrobial activities	• Enhanced scavenger receptors • Enhanced mannose receptors • Tissue repair and regeneration • Anti-inflammatory • Would healing • Immunologic homeostasis • Th2 response

Figure 1.
Polarization profile of placental macrophages. Summary of the molecules involved in polarization profiles inducing the expression of several proteins leading to several functions of placental macrophages.

reappraisal based on the type of agonist. Hence, the concept of polarization should include the source of macrophages, the type of activation and a collection of activation markers. We have proposed to adopt a nomenclature related to the agonist: M(IL-4), M(IFN-γ), M(IL-10), M(GC), M(Ig) and M(LPS) [20].

3. Macrophage polarization during bacterial infections

The M1 profile is classically associated with control of bacterial infections but its definition is variable among publications. In some reports, only few inflammatory mediators (cytokines and inducible nitric oxide synthase (iNOS)) are considered whereas, in others, a combination of markers is used with large sets of genes or proteins [16, 21, 22]. In *in vitro* studies, we and others reported that a M1 profile is found in response to several bacterial pathogens including *Salmonella typhimurium*, *Orientia tsutsugamushi*, *Legionella pneumophila*, *Francisella* spp., *Rickettsia montanensis*, *Shigella dysenteriae*, *Bartonella* spp., *Mycobacterium ulcerans*, *Chlamydia spp.* or *Listeria monocytogenes* [16, 23–27]. The M1 profile is not synonymous of cure of infections since inappropriate M1 response to infection may be deleterious to the host. In animal models of sepsis, M1 phenotype is prevailing in animals that died dye? [28]. This paralysis of immune system may be modelized in models of LPS tolerance. Hence, in repeatedly treated macrophages by LPS, a M2 profile of macrophages becomes prevailing in the late phase of sepsis. The addition of IFN-γ produced by NK cells may reprogram macrophages toward a M1 phenotype [29].

On another hand, bacteria such as *Yersinia spp*. [30], *Ehrlichia muris* [31], *Chlamydia pneumoniae* [32], *Borrelia burgdorferi* [33], *Salmonella typhimurium* [34] or *Rickettsia conorii* [35] favor the occurrence of M2 profiles in macrophages.

As example, we reported that macrophages infiltrating lamina propria during Whipple's disease, an infectious disease due to *Tropheryma whipplei*, are clearly polarized toward M2 phenotype [36]. The M2 profile is a source of a relative consensus and consists of a panel of lectin-like molecules, arginase-1 (Arg1) and a lot of immunoregulatory genes and proteins. It is noteworthy that the number of bacteria inducing a M2 profile is more limited than those inducing a M1 profile. This may be related to the fact that antibacterial responses are of Th1-type rather than Th2-type.

The survival and the replication of pathogenic bacteria within macrophages may rely on strategies interfering with their polarization. *Shigella flexneri* escapes to TLR-4 recognition in murine macrophages via the expression of a truncated form of LPS (hypoacetylated) [37]. This strategy leads to a decreased inflammatory response and prevents the development of M1 response. *Staphylococcus aureus* inhibits NF-κB activity in mice, which is associated with an inhibition of the M1 phenotype of macrophages [38]. This may be related to the resistance of the biofilm of *S. aureus* to macrophage invasion through a decreased expression of inflammatory mediators including IL-1β, TNF, iNOS and an increased expression of Arg1, suggesting a M2 reprogramming [39]. *M. tuberculosis* also interferes with the M1 polarization profile of macrophages by inhibiting phagosome maturation and NF-κB activation [40] or the stimulation of the pathway of Wingless-type MMTV integration site family, member 6 (Wnt6), leading to a M2-like polarization [41]. Interestingly, it was reported that during *M. tuberculosis* infection, macrophage population found in granuloma are mainly TCR^+ that were directly involved in the maintain of the granuloma structure in an TNF-dependent manner [42]. Considered as a distinct subpopulation, macrophage TCR^+ were suggested to present specific characteristics and functionalities whose polarization status is not yet known.

Coxiella burnetii is the cause of Q fever that targets monocytes and macrophages and macrophage polarization may reflect the different steps of disease progression [43]. *C. burnetii* infects monocytes and macrophages, but only M2 polarization environment favorizes their survival [44]. In this context, *C. burnetii* infection leads to a M2 activation of human macrophages including alveolar and monocyte-derived macrophages (MDMs) [16, 45]. This M2 activation is atypical, characterized by the expression of both M2 (TGF-β, CCL18, Arg1, mannose receptor and IL-1 receptor antagonist) and M1 (IL-6 and IL-18) markers. In contrast, *C. burnetii* elicits M1 profile in monocytes in which bacteria do not replicate but only survive [16]. The deficiency of M1 markers, using $NOS2^{-}/^{-}$ or $IFN\gamma^{-}/^{-}$ mice, leads to bacterial replication whereas *C. burnetii* replication is impaired in $IL\text{-}4^{-}/^{-}$ mice [46]. In patients with Q fever, the polarization state of macrophages is closely dependent on the form of Q fever disease including acute or persistent infection. Our team reported the central role of IL-10 associated with uncontrolled *C. burnetii* replication in macrophages from patients with persistent Q fever [47], as well as the bacterial persistence in transgenic mice with IL-10 overexpression in the macrophage compartment [48]. These results suggest that a M2b (IL-10-dependent) profile is associated with bacterial persistence in patients with persistent Q fever.

4. Models of macrophage polarization and methods of study

The evaluation of macrophage polarization depends on cell type (primary cells *versus* cell lines) and origin (murine *versus* human macrophages). The murine (RAW264.7 and J774) and human (THP-1) cell lines have been largely used to study macrophage polarization but the immaturity of murine cell lines limits experimental conclusions. The THP-1 cell line is a robust and proliferative cell line that differentiated into "macrophage-like" following phorbol 12-myristate 13-acetate or M-CSF treatment. In contrast to primary macrophages, THP-1 cells are easily transfected to modify genes involved in polarization pathways. Despite these advantages, the THP-1 cell line presents a lack of physiological relevance and should be dedicated to basic research [49]. Rodents provide a convenient model for macrophage studies since all macrophage compartments are accessible. For a long time, peritoneal macrophages have been the gold standard of macrophage studies despite of their great heterogeneity because they were isolated from peritoneal cavity or from exudates in great quantities. As resident peritoneal macrophages are of M2-type, there may be a concern for their use in polarization experiments [2]. Bone marrow-derived macrophages (BMDMs) can be isolated from wild type and transgenic mice and they represent murine macrophage primary cells mostly used for the investigation of macrophage polarization; these cells have the advantage to present low donor variability and to be genetically modifiable [50].

In healthy humans and patients, primary macrophages derived from peripheral blood mononuclear cells (PBMCs) constitute the most practical model, especially to evaluate the polarization profile. Monocytes are isolated from PBMCs using $CD14^+$ positive selection and differentiated into monocyte-derived macrophages (MDMs) that are not immortalized and do not proliferate. MDMs are produced in large quantities that allow the evaluation of several polarization markers [51]. Nevertheless, there is large donor variability, and cells from certain donors do not respond to polarizing agonists. This variability among individuals may point to *in vitro* cell isolation techniques or artificial differentiation techniques, which could modify transcriptional profile. Recently, it has been showed that macrophages derived from monocytes issued from human stem cells (embryonic or pluripotent) represent a powerful tool to investigate human macrophage polarization [52]. Takata et al. generated human macrophages from induced pluripotent stem cells (iPSCs): these iPSC-derived primitive macrophages (iMacs) exhibit all the criteria of human MDMs [53]. Besides the differentiation ex vivo of monocytes or stem cells into macrophages, the access to tissue macrophages in humans remains a major pitfall. An indirect strategy to reproduce immune response in tissues consists in the formation of granulomas using PBMCs. We showed that granulomatous macrophages share gene expression signature with IFN-γ-stimulated macrophages and thus exhibit a M1 profile [54, 55]. The development of 3D bioengineered tissue model in which macrophages are in their natural environment will be a strategy for future evaluation [56]. Only some tissue macrophage populations are directly available such as alveolar macrophages obtained from bronchoalveolar lavages (BAL). In addition to ethical restriction of BAL in healthy controls, their purity is a concern for investigators, making standardization almost impossible. Placental macrophages are an original population of macrophages of both maternal and fetal origin. We developed a simple method to isolate and characterize them [57]. As placental macrophages are obtained after delivery, the investigation of their polarization is reserved to retrospective studies. Biopsies of pathological tissues are a source of heterogeneous macrophage populations. Hence, we obtained interesting results about M2 polarization of macrophages in intestinal biopsies of patients with Whipple's disease in which the accumulation of macrophages in lamina propria is a clue for the diagnosis but, again, it is not achievable in healthy subjects. Finally, oncologists have a real expertise in macrophage polarization in tumor-associated macrophages (TAMs) [58]. TAMs were found considered as M2-myeloid population in order to maintain a tolerance in the tumor microenvironment [59–61]. They were considered as a marker for recurrence of cancer [62] and their accumulation in tumor microenvironment is associated with a poor prognosis. They presented an ability to switch to M1 phenotype during anti-cancer treatment [63] suggesting that this polarization change could constitute a therapeutic approach [64]. We have to learn lessons from results of polarization in TAMs to translate them to bacterial infections.

The evaluation of macrophage polarization needs a set of markers rather than a single molecule. This is exemplified by the use of a scavenger receptor, CD163, as a prototype of M2 marker. Indeed, CD163 is expressed by M1 cells and non-myeloid cells although at lower level [65]. The same comment can be done with iNOS, a M1 marker, also expressed by endothelial cells and arterial wall smooth muscle cells [66]. The development of high-throughput methods (omics technics) has offered the opportunity to provide convenient sets of polarization markers. The transcriptomics methods such as microarray had been a strategy to investigate macrophage polarization since they provide a large panel of transcripts associated with different modes of polarization (**Figure 1**). Martinez et al. reported transcriptomic analysis of activated macrophages: 5.2% and 0.3% of transcripts are associated with M1 and M2 polarization profiles, respectively [14]. Few years later, Xue et al. performed a transcriptomic analysis of human macrophages stimulated by various panels of agonists [67]. They identified nine specific distinct profiles according the agonist used, and a common transcriptomic signature, which was pertinent to isolate a polarized signature in inflammatory and infectious diseases outside of cancer [68]. Some alternative approaches to microarray such as nanostring method uses directly a panel of genes to measure their variation and may be convenient to investigate macrophage polarization [69]. Whatever the method, gene expression data must be controlled by quantitative RT-PCR, a very sensitive technic that needs low amounts of cells [70]. Discrepancies between microarray and quantitative RT-PCR have been often observed in macrophage polarization studies. The emergence of single cell RNA-sequencing (scRNA-seq) method might provide a powerful tool for analysis macrophage populations including their phenotype and therefore their polarization profile. Interestingly, the used of scRNA-seq permitted to show that M2 macrophages express varying levels of Arg1, challenging the dogma that macrophages with M2 profile all express Arg1 [71].

All these methods measure the expression of genes associated with macrophage polarization. This approach does not have the robustness of methods determining the expression of proteins. The enzyme-linked immunosorbent assay (ELISA) has the disadvantage to measure isolated secreted molecules associated with a given profile. The simultaneous measurement of up to 50 proteins using Luminex assays constitutes an interesting option but the cost and the specialized detection equipment represent a disadvantage. We previously investigated macrophage polarization by a proteomic approach using MALDI-TOF mass spectrometry technique combined with gel electrophoresis [19, 72]. This combined approach allows the determination of M1 signature of human macrophages stimulated with IFN-γ, LPS or bacteria. Moreover, different subtypes of M1 and M2 polarized macrophages have been identified using this approach [72].

The flow cytometry and CyTOF techniques offer a better investigation of macrophage phenotypes through the investigation of protein expression at a single cell resolution level. Hence, CyTOF panels have been proposed to measure polarization markers and, combined with high dimensional analysis, CyTOF enables the identification of novel functional macrophage subsets [73]. The emergence of cycling imaging that purposes to stain cells with different cocktail markers after bleaching allows the detection of more than 30 markers at once [74]. Finally, basic methods such as cell morphology could be used to evaluate functional phenotypes of polarized macrophages [75]. Indeed, polarized M2 murine macrophages are more elongated than M1 cells [76, 77].

A new and exciting field of exploring macrophage polarization, the study of metabolic changes, has recently emerged. LPS ligation of TLR-4 elicits a shift to glycolytic metabolism with impaired mitochondrial respiration. Associated with IFN-γ, LPS induces alterations in tricarboxylic acid cycle. In contrast, IL-4 responses are associated with a shift to oxidative metabolism. Hence, the M2 program associates changes in polyamine synthesis and fatty acid oxidation [78, 79].

This huge diversity of methods exploring macrophage activation including macrophage polarization needs to define the conditions of using these methods and the stratification of indications.

5. Recommendations for measuring macrophage polarization in infected patients

The interest of measuring macrophage polarization in patients is to assess activation status of macrophages to stratify them and to measure their response to treatment. The investigation of macrophage polarization in infected patients requires the choice of pertinent cell types and of the method of measurement. Studies with macrophages from healthy controls stimulated *in vitro* with polarizing ligands are needed to collect specific signatures and to standardize those found in patients. When cells are isolated from infected patients, we have to decipher if they are polarized and which agonist is responsible of such activation profile. As a consequence, each signature should contain several molecules for each polarization category and the determination of these different signatures should be easy to perform in biological laboratories. This means that technics for measuring gene expression such as quantitative RT-PCR, phenotyping membrane or intracellular molecules through flow cytometry or molecule secretion using multiplex ELISA should be privileged. Unfortunately, there is no consensus about the content of polarization signature. Some authors used limited number of molecules known to be associated with polarized status of macrophages, other groups including our used signatures obtained from microarray/RNA sequencing data collection. Hence, the comparison of the studies becomes extremely difficult.

The investigation of patients with bacterial infection is limited by accessibility of biological materials in contrast to cancer in whom tissue biopsies are required for the diagnosis. In practice, circulating monocytes, associated or not with lymphocytes, are the major source of myeloid cells. However, it is uncertain that the M1/M2 polarization of tissue macrophages is also found in circulating monocytes. We compared the polarization of monocytes and MDMs from healthy donors in response to canonical agonists of macrophage M1/M2 polarization, IFN-γ and IL-4. While the two cytokines elicit clear polarized profile in MDMs, a similar polarization is observed in early stimulated monocytes and is lost after 24 h of treatment [80]. This observation may account for numerous discrepancies found in several examples of infectious diseases. While *M. tuberculosis* induces a M2 profile in macrophages *in vitro* [81, 82], the study of gene expression in patients suffering from active tuberculosis revealed a signature in which neutrophil and type I IFN are prominent but did not reveal a polarized profile [83]. We draw similar conclusions from our investigation of patients suffering from Q fever. *C. burnetii* interferes with M1 polarization of macrophages leading to an atypical M2 program [16] but the investigation of circulating monocytes using microarray and quantitative RT-PCR as a confirmation did not reveal a polarization in patients with acute or persistent Q fever [80]. These two examples do not invalidate the use of polarization concept in patients with an infectious disease but underlines the necessity to analyze the data according the type of myeloid cells. In addition, the use of macrophages differentiated from patient monocytes may be biased by the role of M-CSF in the differentiation process that may affect macrophage polarization. It is likely that studying polarization in tissue macrophages such as alveolar macrophages and intestinal macrophages may be more pertinent.

The biopsies are reserved to severe infections or rare infectious diseases in which they are necessary for the diagnosis as in Whipple's disease. The advantage of such approach has been recently illustrated. In patients with tuberculosis who underwent surgical treatment, the investigation of pulmonary biopsies revealed that M2-like polarization was correlated with multidrug resistance [84]. We are suggesting adopting the guidelines used in oncology to characterize TAMs [66]. The polarization of tissue macrophages may be assessed in histological sections either by isolation of cells and *ex vivo* studies or *in situ*. In this later case, immunohistochemistry (IHC) is the best strategy for studying macrophage polarization. The choice of detection method, immunofluorescence or chromogenic method, is discussed. As most samples are fixed with formalin and embedded in paraffin, the chromogenic method is the most convenient. The limit of IHC is the number of available antibodies. Hence, Jayasingam et al. recommend to use double IHC staining: CD68/iNOS or CD68/HLA-DR for M1 macrophages and CD68/CD163 for M2 macrophages. This is in contradiction with the concept of signature and it is necessary to provide new technological solutions to better characterize macrophage polarization in tissues. For instance, mass spectrometry imaging would be useful to analyze macrophages in tissues as already done in tumors. The development of mass cytometry will be interesting for phenotyping tissue macrophages [85, 86].

The concept of macrophage polarization has reached adulthood. If it is extremely efficient for pathophysiological studies, it needs to be adapted to the requirements of clinical investigations. This requires to follow the guidelines we defined several years ago according each type of agonist instead of too imprecise categories such as M1 or M2 cells. It also requires new technical solutions to directly investigate macrophages within tissues. Finally, we have to propose alternatives to biopsy sampling in infected patients who do not require such aggressive procedure.

Acknowledgements

Soraya Mezouar was supported by a “Fondation pour la Recherche Médicale” postdoctoral fellowship (reference: SPF20151234951). This work was supported by the French Government under the “Investissements d'avenir” (Investments for the future) program managed by the “Agence Nationale de la Recherche” (reference: 10-IAHU-03).

Declaration of interest

The authors declare no competing interests.

Author contributions

S.M. and J.L.M. conceived and wrote the manuscript.

Author details

Soraya Mezouar[1,2]* and Jean-Louis Mege[1,2,3]

1 Aix-Marseille University, MEPHI, IRD, APHM, Marseille, France

2 IHU-Mediterranean Infection, Marseille, France

3 UF Immunology Department, APHM, Marseille, France

*Address all correspondence to: soraya.mezouar@univ-amu.fr

References

[1] Tauber AI. Metchnikoff and the phagocytosis theory. Nature Reviews. Molecular Cell Biology. 2003;**4**(11):897-901

[2] Epelman S, Lavine KJ, Randolph GJ. Origin and functions of tissue macrophages. Immunity. 2014;**41**(1):21-35

[3] Wynn TA, Chawla A, Pollard JW. Macrophage biology in development, homeostasis and disease. Nature. 2013;**496**(7446):445-455

[4] Gordon S. Pattern recognition receptors. Cell. 2002;**111**(7):927-930

[5] Murray PJ, Wynn TA. Protective and pathogenic functions of macrophage subsets. Nature Reviews. Immunology. 2011;**11**(11):723-737

[6] Flannagan RS, Cosío G, Grinstein S. Antimicrobial mechanisms of phagocytes and bacterial evasion strategies. Nature Reviews. Microbiology. 2009;7(5):355-366

[7] Weiss G, Schaible UE. Macrophage defense mechanisms against intracellular bacteria. Immunological Reviews. 2015;**264**(1):182-203

[8] Mills CD, Kincaid K, Alt JM, Heilman MJ, Hill AM. M1/M2 macrophages and the Th1/Th2 paradigm. Journal of Immunology. 2000;**164**(12):6166-6173

[9] Mills CD, Ley K. M1 and M2 macrophages: The chicken and the egg of immunity. Journal of Innate Immunity. 2014;**6**(6):716-726

[10] Mosmann TR, Cherwinski H, Bond MW, Giedlin MA, Coffman RL. Two types of murine helper T cell clone. I. Definition according to profiles of lymphokine activities and secreted proteins. Journal of Immunology. 1986;**136**(7):2348-2357

[11] Stein M, Keshav S, Harris N, Gordon S. Interleukin 4 potently enhances murine macrophage mannose receptor activity: A marker of alternative immunologic macrophage activation. The Journal of Experimental Medicine. 1992;**176**(1):287-292

[12] Gordon S. Alternative activation of macrophages. Nature Reviews. Immunology. 2003;**3**(1):23-35

[13] Mantovani A, Sica A, Sozzani S, Allavena P, Vecchi A, Locati M. The chemokine system in diverse forms of macrophage activation and polarization. Trends in Immunology. 2004;**25**(12):677-686

[14] Martinez FO, Gordon S, Locati M, Mantovani A. Transcriptional profiling of the human monocyte-to-macrophage differentiation and polarization: New molecules and patterns of gene expression. Journal of Immunology. 2006;**177**(10):7303-7311

[15] Martinez FO, Helming L, Milde R, Varin A, Melgert BN, Draijer C, et al. Genetic programs expressed in resting and IL-4 alternatively activated mouse and human macrophages: Similarities and differences. Blood. 2013;**121**(9):e57-e69

[16] Benoit M, Desnues B, Mege J-L. Macrophage polarization in bacterial infections. Journal of Immunology. 2008;**181**(6):3733-3739

[17] Martinez FO, Sica A, Mantovani A, Locati M. Macrophage activation and polarization. Frontiers in Bioscience. 2008;**13**:453-461

[18] Mosser DM, Edwards JP. Exploring the full spectrum of macrophage activation. Nature Reviews. Immunology. 2008;**8**(12):958-969

[19] Ouedraogo R, Daumas A, Capo C, Mege J-L, Textoris J. Whole-cell

MALDI-TOF mass spectrometry is an accurate and rapid method to analyze different modes of macrophage activation. Journal of Visualized Experiments. 2013;**82**:50926

[20] Murray PJ, Allen JE, Biswas SK, Fisher EA, Gilroy DW, Goerdt S, et al. Macrophage activation and polarization: Nomenclature and experimental guidelines. Immunity. 2014;**41**(1):14-20

[21] Jenner RG, Young RA. Insights into host responses against pathogens from transcriptional profiling. Nature Reviews. Microbiology. 2005;**3**(4):281-294

[22] Nau GJ, Richmond JFL, Schlesinger A, Jennings EG, Lander ES, Young RA. Human macrophage activation programs induced by bacterial pathogens. Proceedings of the National Academy of Sciences of the United States of America. 2002;**99**(3):1503-1508

[23] Shaughnessy LM, Swanson JA. The role of the activated macrophage in clearing *Listeria monocytogenes* infection. Frontiers in Bioscience. 2007;**12**:2683-2692

[24] Rottenberg ME, Gigliotti-Rothfuchs A, Wigzell H. The role of IFN-γ in the outcome of chlamydial infection. Current Opinion in Immunology. 2002;**14**(4):444-451

[25] Chacón-Salinas R, Serafín-López J, Ramos-Payán R, Méndez-Aragón P, Hernández-Pando R, Van Soolingen D, et al. Differential pattern of cytokine expression by macrophages infected *in vitro* with different *Mycobacterium tuberculosis* genotypes. Clinical and Experimental Immunology. 2005;**140**(3):443-449

[26] Kiszewski AE, Becerril E, Aguilar LD, Kader ITA, Myers W, Portaels F, et al. The local immune response in ulcerative lesions of Buruli disease. Clinical and Experimental Immunology. 2006;**143**(3):445-451

[27] Jouanguy E, Lamhamedi-Cherradi S, Lammas D, Dorman SE, Fondanèche M-C, Dupuis S, et al. A human IFNGR1 small deletion hotspot associated with dominant susceptibility to mycobacterial infection. Nature Genetics. 1999;**21**(4):370-378

[28] Mehta A, Brewington R, Chatterji M, Zoubine M, Kinasewitz GT, Peer GT, et al. Infection-induced of M1 and M2 phenotypes in circulating monocytes: Role in immune monitoring and early prognosis of sepsis. Shock. 2004;**22**(5):423-430

[29] Bellora F, Castriconi R, Dondero A, Reggiardo G, Moretta L, Mantovani A, et al. The interaction of human natural killer cells with either unpolarized or polarized macrophages results in different functional outcomes. Proceedings of the National Academy of Sciences of the United States of America. 2010;**107**(50):21659-21664

[30] Tumitan ARP, Monnazzi LGS, Ghiraldi FR, Cilli EM, Machado de Medeiros BM. Pattern of macrophage activation in *Yersinia*-resistant and *Yersinia*-susceptible strains of mice. Microbiology and Immunology. 2007;**51**(10):1021-1028

[31] Haloul M, Oliveira ERA, Kader M, Wells JZ, Tominello TR, El Andaloussi A, et al. mTORC1-mediated polarization of M1 macrophages and their accumulation in the liver correlate with immunopathology in fatal ehrlichiosis. Scientific Reports. 2019;**9**(1):14050

[32] Buchacher T, Ohradanova-Repic A, Stockinger H, Fischer MB, Weber V. M2 polarization of human macrophages favors survival of the intracellular pathogen *Chlamydia pneumoniae*. PLoS One. 2015;**10**(11):e0143593

[33] Lasky CE, Olson RM, Brown CR. Macrophage polarization during murine *Lyme borreliosis*. Infection and Immunity. 2015;**83**(7):2627-2635

[34] Eisele NA, Ruby T, Jacobson A, Manzanill PS, Cox JS, Lam L, et al. Cell Host & Microbe. 2013;**14**(2):171-182

[35] Curto P, Santa C, Allen P, Manadas B, Simões I, Martinez JJ. A pathogen and a non-pathogen spotted fever group Rickettsia trigger differential proteome signatures in macrophages. Frontiers in Cellular and Infection Microbiology. 2019;**9**:43

[36] Desnues B, Lepidi H, Raoult D, Mege J. Whipple disease: Intestinal infiltrating cells exhibit a transcriptional pattern of M2/alternatively activated macrophages. The Journal of Infectious Diseases. 2005;**192**(9):1642-1646

[37] Paciello I, Silipo A, Lembo-Fazio L, Curcuru L, Zumsteg A, Noel G, et al. Intracellular Shigella remodels its LPS to dampen the innate immune recognition and evade inflammasome activation. Proceedings of the National Academy of Sciences of the United States of America. 2013;**110**(46):E4345-E4354

[38] Xu F, Kang Y, Zhang H, Piao Z, Yin H, Diao R, et al. Akt1-mediated regulation of macrophage polarization in a murine model of *Staphylococcus aureus* pulmonary infection. The Journal of Infectious Diseases. 2013;**208**(3):528-538

[39] Thurlow LR, Hanke ML, Fritz T, Angle A, Aldrich A, Williams SH, et al. *Staphylococcus aureus* biofilms prevent macrophage phagocytosis and attenuate inflammation in vivo. Journal of Immunology. 2011;**186**(11):6585-6596

[40] Lugo-Villarino G. Macrophage polarization: Convergence point targeted by *Mycobacterium tuberculosis* and HIV. Frontiers in Immunology. 2011;**2**:43. Available from: http://journal.frontiersin.org/article/10.3389/fimmu.2011.00043/abstract

[41] Schaale K, Brandenburg J, Kispert A, Leitges M, Ehlers S, Reiling N. Wnt6 is expressed in granulomatous lesions of *Mycobacterium tuberculosis* – infected mice and is involved in macrophage differentiation and proliferation. The Journal of Immunology. 2013;**191**(10):5182-5195

[42] Beham AW, Puellmann K, Laird R, Fuchs T, Streich R, Breysach C, et al. A TNF-regulated recombinatorial macrophage immune receptor implicated in granuloma formation in tuberculosis. PLoS Pathogens. 2011;**7**(11):e1002375

[43] Eldin C, Mélenotte C, Mediannikov O, Ghigo E, Million M, Edouard S, et al. From Q fever to *Coxiella burnetii* infection: A paradigm change. Clinical Microbiology Reviews. 2017;**30**(1):115-190

[44] Amara AB, Bechah Y, Mege J-L. Immune response and *Coxiella burnetii* invasion. Advances in Experimental Medicine and Biology. 2012;**984**:287-298

[45] Dragan AL, Voth DE. *Coxiella burnetii*: International pathogen of mystery. Microbes and Infection. April 2019;**22**(3):100-110

[46] Fernandes TD, Cunha LD, Ribeiro JM, Massis LM, Lima-Junior DS, Newton HJ, et al. Murine alveolar macrophages are highly susceptible to replication of *Coxiella burnetii* phase II *in vitro*. Infection and Immunity. 2016;**84**(9):2439-2448

[47] Honstettre A, Imbert G, Ghigo E, Gouriet F, Capo C, Raoult D, et al. Dysregulation of cytokines in acute Q fever: Role of interleukin-10 and tumor necrosis factor in chronic evolution of Q fever. Journal of Infectious Diseases. 2003;**187**(6):956-962

[48] Meghari S, Bechah Y, Capo C, Lepidi H, Raoult D, Murray PJ, et al. Persistent *Coxiella burnetii* infection in mice overexpressing IL-10: An efficient model for chronic Q fever pathogenesis. PLoS Pathogens. 2008;**4**(2):e23

[49] Baxter EW, Graham AE, Re NA, Carr IM, Robinson JI, Mackie SL, et al. Standardized protocols for differentiation of THP-1 cells to macrophages with distinct M(IFNγ + LPS), M(IL-4) and M(IL-10) phenotypes. Journal of Immunological Methods. 2020;**478**:112721

[50] Pineda-Torra I, Gage M, de Juan A, Pello OM. Isolation, culture, and polarization of murine bone marrow-derived and peritoneal macrophages. Methods in Molecular Biology. 2015;**1339**:101-109

[51] Jin X, Kruth HS. Culture of macrophage colony-stimulating factor differentiated human monocyte-derived macrophages. Journal of Visualized Experiments. 2016;**112**:54244

[52] van Wilgenburg B, Browne C, Vowles J, Cowley SA. Efficient, long term production of monocyte-derived macrophages from human pluripotent stem cells under partly-defined and fully-defined conditions. PLoS One. 2013;**8**(8):e71098

[53] Takata K, Kozaki T, Lee CZW, Thion MS, Otsuka M, Lim S, et al. Induced-pluripotent-stem-cell-derived primitive macrophages provide a platform for modeling tissue-resident macrophage differentiation and function. Immunity. 2017;**47**(1):183-198.e6

[54] Faugaret D, Ben Amara A, Alingrin J, Daumas A, Delaby A, Lépolard C, et al. Granulomatous response to *Coxiella burnetii*, the agent of Q fever: The lessons from gene expression analysis. Frontiers in Cellular and Infection Microbiology. 15 December 2014;**4**:172. Available from: http://journal.frontiersin.org/article/10.3389/fcimb.2014.00172/abstract

[55] Delaby A, Gorvel L, Espinosa L, Lépolard C, Raoult D, Ghigo E, et al. Defective monocyte dynamics in Q fever granuloma deficiency. The Journal of Infectious Diseases. 2012;**205**(7):1086-1094

[56] Roh TT, Chen Y, Paul HT, Guo C, Kaplan DL. 3D bioengineered tissue model of the large intestine to study inflammatory bowel disease. Biomaterials. 2019;**225**:119517

[57] Mezouar S, Ben Amara A, Chartier C, Gorvel L, Mege J-L. A fast and reliable method to isolate human placental macrophages. Current Protocols in Immunology. 2019;**125**(1):e77

[58] Tamura R, Tanaka T, Yamamoto Y, Akasaki Y, Sasaki H. Dual role of macrophage in tumor immunity. Immunotherapy. 2018;**10**(10):899-909

[59] Mantovani A, Schioppa T, Porta C, Allavena P, Sica A. Role of tumor-associated macrophages in tumor progression and invasion. Cancer Metastasis Reviews. 2006;**25**(3):315-322

[60] Sica A, Schioppa T, Mantovani A, Allavena P. Tumour-associated macrophages are a distinct M2 polarized population promoting tumour progression: Potential targets of anti-cancer therapy. European Journal of Cancer. 2006;**42**(6):717-727

[61] Allavena P, Sica A, Garlanda C, Mantovani A. The Yin-Yang of tumor-associated macrophages in neoplastic progression and immune surveillance. Immunological Reviews. 2008;**222**(1):155-161

[62] Suriano F, Santini D, Perrone G, Amato M, Vincenzi B, Tonini G, et al. Tumor associated macrophages

polarization dictates the efficacy of BCG instillation in non-muscle invasive urothelial bladder cancer. Journal of Experimental & Clinical Cancer Research. 2013;**32**(1):87

[63] Chen D, Xie J, Fiskesund R, Dong W, Liang X, Lv J, et al. Chloroquine modulates antitumor immune response by resetting tumor-associated macrophages toward M1 phenotype. Nature Communications. 2018;**9**(1):873

[64] Mantovani A, Germano G, Marchesi F, Locatelli M, Biswas SK. Cancer-promoting tumor-associated macrophages: New vistas and open questions. European Journal of Immunology. 2011;**41**(9):2522-2525

[65] Hu JM, Liu K, Liu JH, Jiang XL, Wang XL, Chen YZ, et al. CD163 as a marker of M2 macrophage, contribute to predicte aggressiveness and prognosis of Kazakh esophageal squamous cell carcinoma. Oncotarget. 2017;**8**(13):21526-21538

[66] Jayasingam SD, Citartan M, Thang TH, Mat Zin AA, Ang KC, Ch'ng ES. Evaluating the polarization of tumor-associated macrophages into M1 and M2 phenotypes in human cancer tissue: Technicalities and challenges in routine clinical practice. Frontiers in Oncology. 2020;**9**:1512

[67] Xue J, Schmidt SV, Sander J, Draffehn A, Krebs W, Quester I, et al. Transcriptome-based network analysis reveals a spectrum model of human macrophage activation. Immunity. 2014;**40**(2):274-288

[68] Hussell T, Bell TJ. Alveolar macrophages: Plasticity in a tissue-specific context. Nature Reviews. Immunology. 2014;**14**(2):81-93

[69] Ayaub EA, Tandon K, Padwal M, Imani J, Patel H, Dubey A, et al. IL-6 mediates ER expansion during hyperpolarization of alternatively activated macrophages. Immunology and Cell Biology. 2019;**97**(2):203-217

[70] Spiller KL, Wrona EA, Romero-Torres S, Pallotta I, Graney PL, Witherel CE, et al. Differential gene expression in human, murine, and cell line-derived macrophages upon polarization. Experimental Cell Research. 2016;**347**(1):1-13

[71] Arlauckas SP, Garren SB, Garris CS, Kohler RH, Oh J, Pittet MJ, et al. Arg1 expression defines immunosuppressive subsets of tumor-associated macrophages. Theranostics. 2018;**8**(21):5842-5854

[72] Ouedraogo R, Flaudrops C, Ben Amara A, Capo C, Raoult D, Mege J-L. Global analysis of circulating immune cells by matrix-assisted laser desorption ionization time-of-flight mass spectrometry. PLoS One. 2010;**5**(10):e13691

[73] Roussel M, Bartkowiak T, Irish JM. Picturing polarized myeloid phagocytes and regulatory cells by mass cytometry. Mass Cytometry. 2019;**1989**:217-226. DOI: 10.1007/978-1-4939-9454-0_14

[74] Giedt RJ, Pathania D, Carlson JCT, McFarland PJ, del Castillo AF, Juric D, et al. Single-cell barcode analysis provides a rapid readout of cellular signaling pathways in clinical specimens. Nature Communications. 2018;**9**(1):4550

[75] Marklein RA, Lam J, Guvendiren M, Sung KE, Bauer SR. Functionally-relevant morphological profiling: A tool to assess cellular heterogeneity. Trends in Biotechnology. 2018;**36**(1):105-118

[76] Rodell CB, Arlauckas SP, Cuccarese MF, Garris CS, Li R, Ahmed MS, et al. TLR7/8-agonist-loaded nanoparticles promote the polarization of tumour-associated

macrophages to enhance cancer immunotherapy. Nature Biomedical Engineering. 2018;**2**:578-588

[77] Nishio M, Urakawa N, Shigeoka M, Takase N, Ichihara Y, Arai N, et al. Software-assisted morphometric and phenotype analyses of human peripheral blood monocyte-derived macrophages induced by a microenvironment model of human esophageal squamous cell carcinoma: Image analysis of human PBMC-derived macrophages. Pathology International. 2016;**66**(2):83-93

[78] Phan AT, Goldrath AW, Glass CK. Metabolic and epigenetic coordination of T cell and macrophage immunity. Immunity. 2017;**46**(5):714-729

[79] Caputa G, Castoldi A, Pearce EJ. Metabolic adaptations of tissue-resident immune cells. Nature Immunology. 2019;**20**(7):793-801

[80] Mehraj V, Textoris J, Ben Amara A, Ghigo E, Raoult D, Capo C, et al. Monocyte responses in the context of Q fever: From a static polarized model to a kinetic model of activation. The Journal of Infectious Diseases. 2013;**208**(6):942-951

[81] Huang Z, Luo Q, Guo Y, Chen J, Xiong G, Peng Y, et al. *Mycobacterium tuberculosis*-induced polarization of human macrophage orchestrates the formation and development of tuberculous granulomas *in vitro*. PLoS ONE. 2015;**10**(6):e0129744

[82] Roy S, Schmeier S, Kaczkowski B, Arner E, Alam T, Ozturk M, et al. Transcriptional landscape of *Mycobacterium tuberculosis* infection in macrophages. Scientific Reports. 2018;**8**(1):6758

[83] Berry MPR, Graham CM, McNab FW, Xu Z, Bloch SAA, Oni T, et al. An interferon-inducible neutrophil-driven blood transcriptional signature in human tuberculosis. Nature. 2010;**466**(7309):973-977

[84] Cho HJ, Lim Y-J, Kim J, Koh W-J, Song C-H, Kang M-W. Different macrophage polarization between drug-susceptible and multidrug-resistant pulmonary tuberculosis. BMC Infectious Diseases. 2020;**20**(1):81

[85] Maby P, Corneau A, Galon J. Phenotyping of tumor infiltrating immune cells using mass-cytometry (CyTOF). Methods in Enzymology. 2020;**632**:339-368

[86] Mrdjen D, Pavlovic A, Hartmann FJ, Schreiner B, Utz SG, Leung BP, et al. High-dimensional single-cell mapping of central nervous system immune cells reveals distinct myeloid subsets in health, aging, and disease. Immunity. 2018;**48**(2):380-395.e6

Section 3

Macrophage for Neuro-Muscular Disease

Chapter 5

Pathogenic Role of iNOs+ M1 Effector Macrophages in Fibromyalgia

Vishwas Tripathi, Amaresh Mishra, Yamini Pathak, Aklank Jain and Hridayesh Prakash

Abstract

Fibromyalgia (FM) or Fibromyalgia Syndrome (FMS) is a neurodegenerative disorder causing musculoskeletal pain, tenderness, stiffness, fatigue, and sleep disorder in the body. It is one of the most common chronic pain conditions, affecting about 6% of the world population. Being refractory, till date, no specific treatment of this disease is available. Accumulating evidences over the last few decades indicate that proinflammatory macrophages, cytokines, & chemokines as the key players in this disease. Recent findings suggest activation of Microglial cells and associated pro-inflammatory signals as one of the major causes of chronic pain in patients suffering from fibromyalgia. Increased density of iNOs/CD68+ M1 effector macrophages has been associated with neuropathic pain models. In light of this, depletion of these pro-inflammatory macrophages has been shown to reduce sensitivity to neuropathic pain. On the other hand, modulating pattern of AGEs (Advanced Glycation End-Products) can also contribute to inactivation of macrophages. These findings strongly suggest that macrophages are critical in both inflammatory and neuropathic pain. Therefore, this chapter highlights the impact of macrophage plasticity in various immunopathological aspects of fibromyalgia.

Keywords: fibromyalgia, Th1/Th2 immune response, M1/M2 macrophages, neurodegeneration

1. Introduction

Fibromyalgia has been considered a rheumatologic disease, also known as fibrositis and myofascial pain syndrome. Fibromyalgia affects the muscles, ligaments & tendons, and bones with no signs of inflammation of the tissue [1]. The origin of fibromyalgia is still not clear, although several hypotheses stated fibromyalgia condition associated with depression and brain-based neuronal dysfunctions i.e. coaxially linked with non-uniform signal transduction mechanism. Preclinical studies addressing the symptoms of fibromyalgia are increased levels of substance P (SP) in the cerebrospinal fluid (CSF) of the individuals. SP is a peptide which is composed of 11 amino acids and acts as a neurotransmitter. It plays a significant role in pain stimulations from the peripheral nervous system to the central nervous system [2]. Fibromyalgia patients showed a 3-fold increase in the levels of substance P in CSF which possibly activates neurokinin (NK) receptors that induce chronic pain [3, 4].

Moreover, besides NK receptors, the excitation of amino acid receptors such as N-methyl D-aspartate (NMDA) receptors also leads to hyperalgesia in fibromyalgia [5]. This is associated with the deficiency of Dopamine during chronic pain in fibromyalgia [6]. It is a neurotransmitter of the Central nervous system (CNS) and regulates the pain processing of CNS. On the other hand, several studies indicating that macrophages, cytokines/chemokines, and oxidative stress are the key mediator of immune activation and inflammation in fibromyalgia condition [7].

Nitric oxide (NO), is a principal determinant of normal endothelial and vascular function. During inflammatory reactions, NO production increases considerably and, contributes to oxidative stress together with other reactive oxygen species (ROS) [8]. Macrophages that have a pro-inflammatory role are called classically-activated (M1) macrophages. Classically-activated (M1) macrophages are activated by Lipopolysaccharide (LPS) and Interferon-gamma (IFN-γ). The role of activated M1 macrophages is to secrete pro-inflammatory cytokines and chemokines and present antigens [9, 10]. Cytokines produced by T helper type 1 (Th1) cells, induce the differentiation of classically activated (M1) macrophages [11, 12]. Some of the pro-inflammatory cytokines including TNF-α, IL-1, IL-6, and IL-8 have been reportedly linked with the immunopathology of fibromyalgia. Prolonged activation of M1 macrophages has been reported to encourage neuro-inflammation which may responsible for the pathogenesis of neuropathic pain among the fibromyalgia patients [13–15]. There is no successful medication yet that has been proven to treat fibromyalgia completely.

1.1 Epidemiology

Fibromyalgia prevalence is more common among women as compared to men and risk increases significantly after growing age [16]. Apparently, in population studies indicate that there is a need for a standardized gender-based diagnostic approach that can allow for more reliable diagnosis and some of the gender biases that relate to women suffering the most. It is estimated globally that there is about 6% of the patient that suffers from this chronic disease out of which 68 percent are females [17]. Patients who need tertiary care pain clinic help, more than 40% outcome measure is a resolution of symptoms of fibromyalgia [18]. Patients with existent chronic rheumatic diseases are having high risk of fibromyalgia (**Figure 1**).

1.2 Etiology and pathophysiology

Fibromyalgia is a chronic pain disorder but the etiology of the disease still not clear [19, 20]. Fibromyalgia is triggered by multiple physical and/or emotional stress factors. There is increasing evidence that revealed the role of macrophages including activated M1 macrophages, and inducible nitric oxide synthase (iNOS) in pain conditions in fibromyalgia [21]. Significantly, macrophages can mediate microglial activation through the production of pro-inflammatory cytokines such as IL-1β, IL-6, and TNF-α [22]. In addition, levels of pro-inflammatory cytokines and chemokines are enhanced in serum and could contribute to inflammation at the systemic level. On the other hand, alteration of central nervous system (CNS) cells occurs in fibromyalgia cause discomfort and sensory perception [5]. The functional neuroimaging technique rests crucially on the identification of pain-sensitive areas in regions of the Brain. Furthermore, differences in activation of pain-sensitive areas of the brain by functional neuroimaging techniques have been revealed in fibromyalgia condition. Several pharmacological studies have shown a genetic predisposition for fibromyalgia though there is no documentation of a definitive candidate gene [23]. About one-third of fibromyalgia patients have a close relative with rheumatic disease/fibromyalgia history (**Figure 2**) [24].

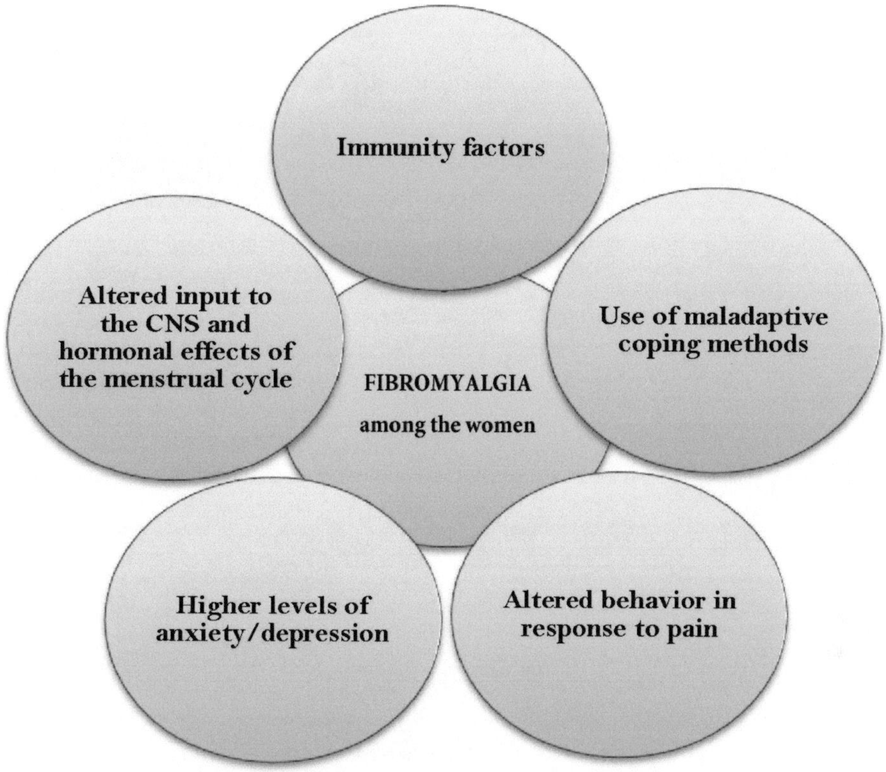

Figure 1.
Reason of high prevalence rate of fibromyalgia among women.

1.2.1 Psychological stress and trauma

Research has also shown that stress is linked with the increasing risk of developing fibromyalgia [25]. Psychologic factors including stress, trauma, anxiety, and depression have been shown their role in pain severity in fibromyalgia patients [26, 27]. Corticotrophin-releasing hormone (CRH), is a well-known hormone for stress response mediators, was found high in the cerebrospinal fluid (CSF) among fibromyalgia patients and was linked with pain [28]. Fibromyalgiais quite common in individuals with mastocytosis [29] where mast cells influence the infiltration, in situ differentiation and inflammatory response of macrophages and neutrophils; in various organs. This type of rapid proliferation of mast cells causes itchy bumps on the skin, diarrhea, and bone pain. Emotional stress is the primary symptom among the individuals suffering from mastocytosis and reportedly found high serum levels of CRH in mastocytosis patients [30]. CRH, nerve growth factor, neurotensin and substance P are known as stress peptides which released in peripheral tissues like blood vessels, muscles, and skin during allergic, immune, and stress reaction in the body. In another study, upregulation of nerve growth factor in the CSF among the patients suffering from fibromyalgia [31] and has been considered as a target for analgesic therapy [32].

1.2.2 Neuro-inflammation

A correlation between macrophages and mast cells (MCs) has been revealed in fibromyalgia [33, 34]. Numerous studies suggested macrophages and MCs play a key role during the pain and inflammation [35–38]. Macrophages and MCs also

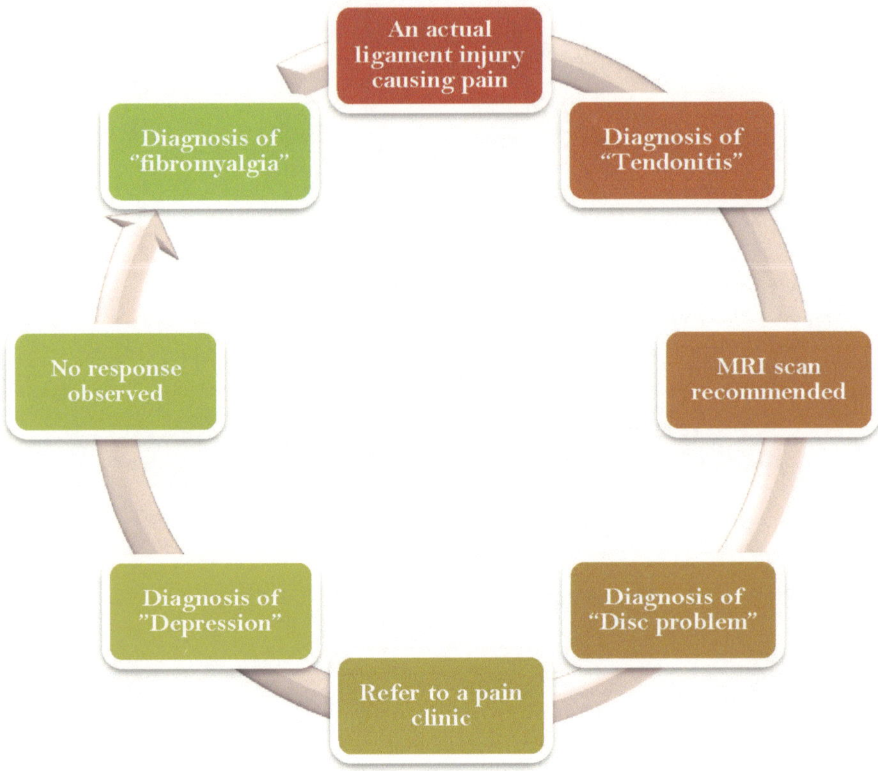

Figure 2.
Progression and diagnosis process of fibromyalgia.

release pro-inflammatory and neuro-sensitizing molecules such as cytokines and chemokines which act as modulators of nociception, and elevated pain sensitivity through their receptors [39–41]. In addition, a growing body of evidence indicated the increased levels of the pro-inflammatory chemokines in both serum and CSF of fibromyalgia patients [42–44]. CSF and IL-17 are also associated with pain, depression, and anxiety which are the key symptoms among the individuals suffering from fibromyalgia [45, 46].

1.2.3 Central sensitization

By far the most important thing to understand about the pain and fatigue induced by fibromyalgia is central pain sensitization. Many nociceptive dorsal root ganglion (DRG) neurons express pro-inflammatory cytokine and chemokine receptors that are upregulated after nerve injury [47]. Long-lasting neuro-inflammation through the upregulation of inflammatory molecules can contribute to the ectopic discharge of sensory neurons, resulting in peripheral sensitization. Prolonged abnormal transmission of pain signaling due to peripheral sensitization triggers central sensitization [48–50], mediated by pain-processing neurons and activation of glial cells (**Figure 3**) [51–54]. Glial cells are activated by various neurotransmitters, such as cytokines, chemokines, and nucleotides and these activated glial cells directly or indirectly involved in central sensitization [10, 55–58]. Moreover, these cells can contribute to brain inflammation and pathogenesis of different brain disorders [59–65].

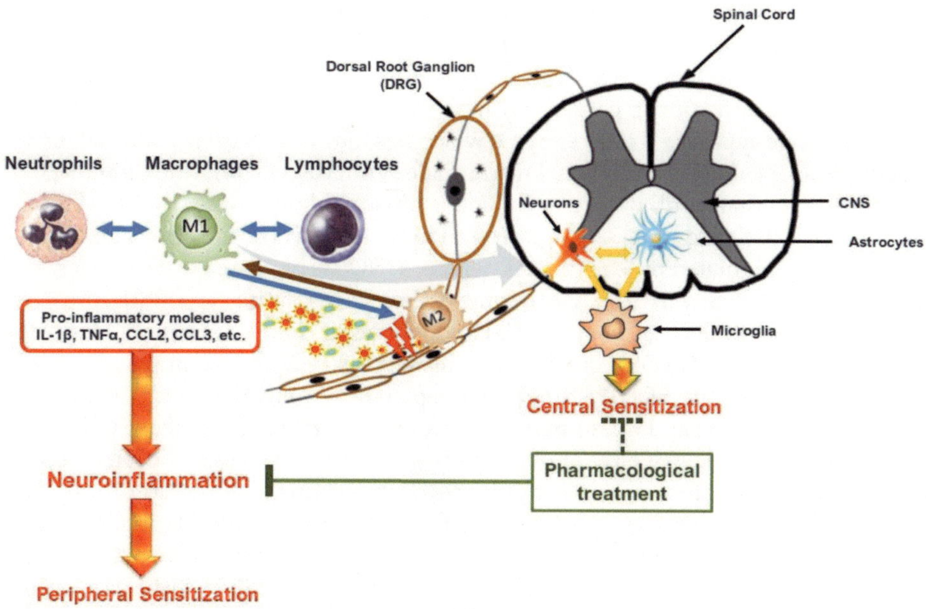

Figure 3.
Contribution of macrophage derived inflammatory response in neuropathic pain in peripheral nervous system.

1.2.4 Fascia

It has been hypothesized that inflammation of fascia is a secondary source of the increased pain transmitting to the spinal cord. Fascia is the thick connective tissue surrounding the cells of the muscles etc. The fascial system covers every part of a muscle and is a thick gel of ground material that suspends muscle cells and fibers [66]. Additionally, muscle innervation is found primarily in the fascia, hence the fascia is highly susceptible. Muscle biopsy, cell studies of FM patients result in higher levels of collagen and symptoms of oxidative stress and tissue damage, indicating fascial inflammation. Although findings were not consistent in considering the hypothesis of fascial inflammation as the fibromyalgia etiology [67]. Conversely, these inflammations may be due to low growth hormone production and HPA axis dysfunction, resulting in increased nociception, central sensitization, and chronic pain [68].

1.2.5 Altered biochemistry

Substance P (SP) is an 11-amino acid peptide IS primarily responsible in the neurotransmission of pain to the central nervous system (CNS). Substance P (SP) is associated with chronic pain found to be elevated with fibromyalgia in the cerebral spinal fluid (CSF) compared to controls [4, 5]. However, the diagnostic study also revealed low neurotransmitter levels involved in regulating sensory perception and also inhibit pain transmission, such as serotonin, norepinephrine, and dopamine, etc.

1.2.6 Nitric oxide synthase (NOS)

Fibromyalgia patients have higher oxidative stress index and lower total nitrite levels than healthy controls [69, 70]. NO plays a crucial role in chronic pain states with cyclooxygenase-2 (COX-2) as termed in central sensitization [71]. Numerous studies considered NO as an important neurotransmitter involved in the pain and sensitization pathways related to NOS activation. Therefore, the NO can actively participate in the hyper sensitization in patients with fibromyalgia. The NO is synthesized by the nitric oxide synthase (NOS) enzyme which has four isoforms i.e. neuronal nitric oxide synthase (nNOS), endothelial nitric oxide synthase (eNOS), mitochondrial nitric oxide synthase (mtNOS), and inducible nitric oxide synthase (iNOS) [72, 73]. The nNOS, eNOS, and mtNOS are expressed in the majority of the cells. The nNOS and eNOS are regulated by Ca2+ fluxes, whereas iNOS is regulated by cytokines. iNOS is expressed only in response to some pathological stimuli typically bypro-inflammatory cytokines and/or bacterial lipopolysaccharide (LPS) [74, 75]. Inducible nitric oxide synthase (iNOS) together with oxidative stress plays an important role in the development of vascular dysfunction in sepsis. In fibromyalgia, a key mediator of immune activation and inflammation is inducible nitric oxide synthase (iNOS), which produces nitric oxide (NO) [8]. Various studies have revealed the iNOS role in the development of inflammatory and neuropathic pain including fibromyalgia. Nitric oxide (NO) has various physiological functions such as vasodilation, muscle relaxation, learning, memory, neurotransmission, several degenerative processes and inflammation [8]. Copious amount of NO production is critical for the inflammatory response and the innate immune system. Overexpression or dysregulation of iNOS is linked with local inflammatory reactions and contributed to various human diseases [75, 76]. Considering this, iNOS inhibitory therapeutics could be promising for the treatment of neurodegenerative pain including fibromyalgia.

2. Role of activated macrophages/myeloid cells in the neuromyalgia

In turn, mast cells interplay with microglia, which are the resident macrophages of the central nervous system that may contribute to increased inflammation through the secretion of cytokines [59, 77]. The cytoplasmic receptor Nod-like receptor-2 (NOD2), and its adaptor-signaling molecule RIPK2, have been shown to be involved in the development of neuropathic pain after peripheral nerve injury. The activation of NOD2 signaling in peripheral macrophage mediates the

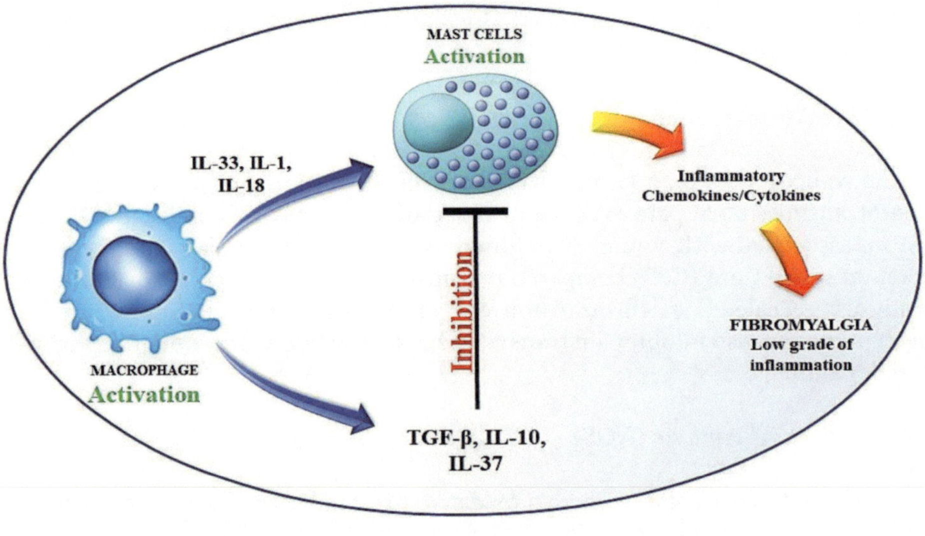

Figure 4.
Pathogenic role of Th1 primed (iNOS+) macrophages and mast cells during progression of fibromyalgia.

development of neuropathic pain through the production of a wide of pro-nociceptive cytokines. The studies strongly suggest the undetermined significance of NOD2 signaling in the development of neuropathic pain and to highlight potential new means of the target for preventing neuropathic pain [78]. Macrophages are important participants in regulating neuro- inflammation (**Figure 4**); consequently, they are considered to be a common peripheral regulator of neuropathic pain [79–81].

3. Current therapies

There is no effective therapeutic approach available for fibromyalgia, but many drugs have been available to reduce its symptoms and pain. Patients suffering from fibromyalgia should integrate pharmacologic therapy with non-pharmacologic therapies [82]. In a study, it has been revealed that multi-component treatment could be effective in the short term for improving key symptoms of fibromyalgia including pain, fatigue, depression, and quality of life [83]. Modern medicine has most certainly come a long way in providing relief from the conditions and diseases of the day, sometimes other options can be just as helpful, if not more beneficial, in providing relief of fibromyalgia symptoms. Moreover, the prescription medications that are commonly used for the treatment of fibromyalgia symptoms can cause negative side effects that the individual must then deal with in addition to the problems along with symptoms of fibromyalgia. In order to regulate inflammation, role of macrophages is very crucial. As we described previously also, macrophages are important to sense damage to the tissue and initiate the recruitment of circulating leukocytes through triggering the chemokines secretion. The direct physical interaction stimulates production of reactive oxygen species (ROS) at the site of the injury. At the late stages of muscle regeneration, macrophages refrain the expression of both pro-inflammatory and anti-inflammatory cytokines and turned to a silenced mode. Conversely, interleukin 4 (IL-4) actions are considered to be regulated by the inhibition of pro inflammatory mediators. In a study, IL-4 blood levels were found reduced among the patients suffering with chronic widespread pain including fibromyalgia when compared with controls [84, 85]. Several study indicating that the endogenous opioid system is essential to the actions of IL-4 and M2 macrophages in pain control [21]. Concluding this, macrophages play a central role in the regulation of inflammation from the beginning to the end. This is a timely area of research to explore the role of macrophages particularly M2 macrophages which sounds more promising for tackling pathological pain.

However, there are some alternative therapeutic approaches to fibromyalgia that have been proven by research to aid in providing relief from its symptoms (**Figure 5**).

3.1 Non-pharmacologic therapy

Patients diagnosed with fibromyalgia must know their illness before starting their medications [86–88]. It was found that educational intervention had significantly better improvement among fibromyalgia patients [82]. In another study, fibromyalgia patients reduce the fear of pain and fear of disease complications using cognitive behavioral therapy [83]. Cardio exercise is suggested for fibromyalgia patients as it helps to improve the sleep and reduce the pain as well [89, 90]. In a study, they uses Chinese stress reduction exercise programs and improvement reported among the fibromyalgia patients to reduce the key symptoms of fibromyalgia [91, 92]. Nutritional supplementation is often used in fibromyalgia, but the objective findings are limited [93, 94]. Coenzyme Q_{10}supplementation in fibromyalgia patients improved the disease symptoms as it reduces the oxidative damage that

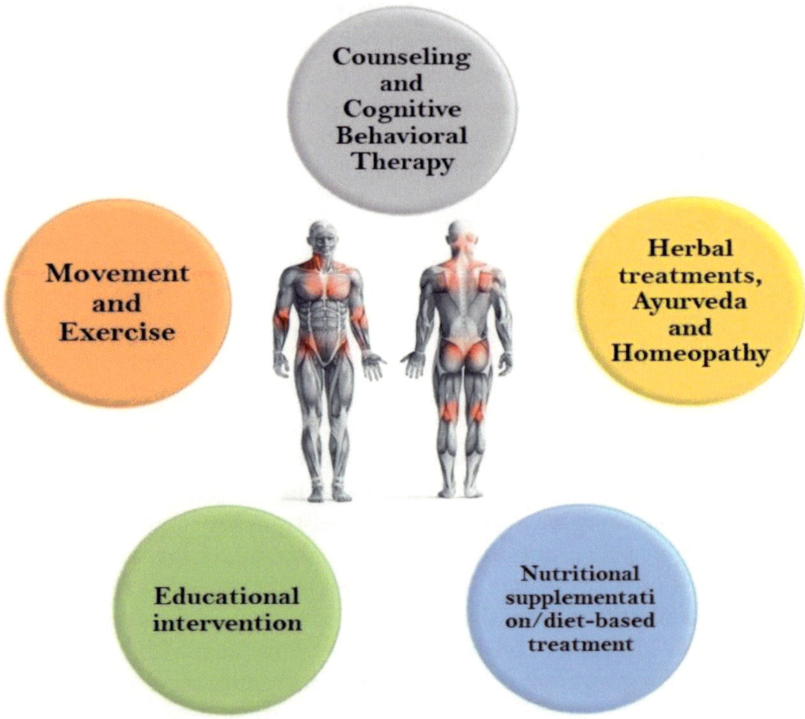

Figure 5.
Effective alternative therapeutic approaches suggested for fibromyalgia patients.

leads to muscle fatigue [95, 96]. Another study revealed that500mg L-carnitine for 20 days has significant benefits to fibromyalgia patients which lasts up to 10 weeks [97]. Natural flavonoids like quercetin and luteolin giving promising results to reduce the key symptoms of fibromyalgia due to its anti-inflammatory, antioxidant, and anti-allergic property [98–101]. Moreover, flavonoids have been discussed as a possible treatment of central nervous system disorders [102, 103].

3.2 Pharmacologic therapy

There's no complete cure available for fibromyalgia, but there are many medicines available to treat fibromyalgia symptoms. Some drugs ease aches, fatigue, and pains, while others may boost your energy or improve your sleep. Fibromyalgia drugs mainly target pain modulatory mechanisms. There is an urgent need to develop new drugs targeting the fibromyalgia mechanism and treat its symptoms (**Table 1**).

4. Immune mediated therapeutic interventions

It is a well-established concept that the immune system plays a crucial role in various chronic pain conditions including fibromyalgia. The immune system involves the release of autoantibodies, pro-inflammatory cytokines, chemokines, substance P, histamine, tumor necrosis factor, interleukins, and prostaglandins [9]. In a study, IL-8 level elevated in the serum of patients suffering from fibromyalgia confirming the relation between fibromyalgia and higher levels of pro-inflammatory cytokines [136]. In another study, the role of the NLRP3 inflammasome in

S.N.	Drug type	Drugs name	Effects on fibromyalgia	Side-effects	References
1.	Antidepressants	Amitriptyline Citalopram,Escitalopram, Fluvoxamine, Fluoxetine, Paroxetine, Sertraline	Improvement in pain, fatigue, and sleep	Drowsiness, Weight gain, Nausea fatigue, Dry mouth, Blurred vision, Constipation, Dizziness, and Change in appetite	[104–107]
2.	Anti-Seizure Medicines	Pregabalin, Gabapentin	Improvement in pain, fatigue, and sleep	Blurry vision, dizziness, Drowsiness, Weight gain, and Swelling of hands/feet	[18, 28, 83, 87, 104, 108–111]
3.	Pain Relievers	acetaminophen and Non-steroidal anti-inflammatory drugs (NSAIDs) like aspirin, ibuprofen, naproxen, and tramadol	Improvement in aches and pains	Heart attack, Stroke ulcers & bleeding in the stomach, Intestines liver damage, Stomach pain, Constipation, Nausea, and Trouble concentrating	[112]
4.	Muscle Relaxants	Cyclobenzaprine (Flexeril) Tizanidine (Zanaflex)	Improvement in pain, fatigue, and sleep	Dry mouth, Dizziness, Blurry vision, Headaches, Chest pain, Nausea, and Fever	[113]
5.	Serotonin inhibitors	Duloxetine, Milnacipran, Reboxetine, Esreboxetine, Citalopram, escitalopram, fluoxetine, paroxetine	Improvement in pain, and depression	Difficulty in sleeping, Headaches, Dizziness, Blurry vision,Constipation/diarrhea, Nausea/vomiting, Dry mouth, and sweating	[104, 105, 114–121]
6.	Gabapentinoid	Pregabalin, gabapentin, Lacosamide	Improvement in pain, fatigue, and sleep	Abnormal eye movements (continuous, & uncontrolled, rolling), Clumsiness, Constipation/ diarrhea, Difficulty speaking, Tiredness, Dry mouth, and Nausea.	[93, 104, 105, 109, 121–124]
7.	Cannabinoid	Nabilone, Dronabinol	Improvement in pain, fatigue, anxiety, and sleep.	Dizziness, Drowsiness, Dry mouth, Feeling "high," Lightheadedness, Headache, and Insomnia	[125–128]
8.	NMDA antagonist	Ketamine	Improvement in pain	High/Low blood pressure, Increased cardiac output, Visual hallucinations, Vivid dreams, and Double vision	[129–132]
9.	Nitrogen-containing bases inhibitors	Methotrexate, Azathioprine, Leflunomide	Improvement in pain	Dizziness, Headache, Tender gums, Decreased appetite, Reddened Eyes, and Hair Loss	[133]
10.	Tumor Necrosis Factor (TNF) inhibitors	Hydroxychloroquine, Adalimumab, Golimumab, Certolizumab, Infliximab, Sulfasalzine and Etanercept	Improvement in pain, and fatigue	Swelling, Redness or itchy skin where your injection was given, A mild nose, throat or sinus infection,Headache, Stomach pain, and Dizziness	[134, 135]

Table 1.
Available pharmacological interventions for fibromyalgia.

fibromyalgia patients along with animal models was investigated. In the outcome of the same study, it has been revealed that increased levels of IL-1b were positively linked with pain in both mice and fibromyalgia patients [137]. This was the first of its kind study to show the relation between inflammasome and increased pro-inflammatory cytokines among the fibromyalgia patients which confirm the direct link between inflammation and pain. Naltrexone and naloxoneisan antagonist of mu-opioid receptors and both were effective to inhibit cytokine expression [138, 139]. Using neuron–glia co-cultures pre-treated with naloxone and subsequently treated with LPS, it was demonstrated that naloxone protects against lipopolysaccharide (LPS)-induced neurotoxicity through the inhibition of the proinflammatory factors and free radicals [139]. Similarly, naltrexone also blocked LPS-induced inflammation and microglial activity and inhibited TNF-α production [139, 140]. Due to the promising preclinical data, clinical trial was conducted for the naltrexone and it has been found that it effectively reduced the key disease symptoms among the fibromyalgia patients [141]. Observational studies provide evidence that vagus nerve activation can down-regulated inflammation through nAChR-mediated inhibition of macrophage function [142–145]. Treatment with nicotine can inhibit the development of inflammatory cytokines release by LPS-stimulated macrophages through the activation of α7 nAChR signaling [143, 144, 146–148]. Following this, anti-inflammatory treatments could be promising in the therapeutics in fibromyalgia condition.

5. Conclusion

Fibromyalgia condition is defined by widespread musculoskeletal pain followed by fatigue, sleep, depression, and anxiety. The available allopathic medicines have their limitations and side effects and to date no permanent treatment of fibromyalgia is available. However, research is going on to find out more alternative options that can be used for treating various chronic conditions. Hence the need of the hour is to explore various alternative therapies for this condition. Despite various research and findings on fibromyalgia, it is found by most of the scientist that this disease is characterized by the abnormal nerve to signal transduction thus “the hypothesis of pain in the brain” is proven here and therefore effective neuronal diagnosis should be conducted before designing the treatment protocols. Thus, the emphasis of treatment must be in-combat with the brain to muscle co-ordinations. It has been well established that the immune system is an important part and plays a key role in the complex pathogenesis of fibromyalgia. Macrophages, inflammatory cytokines, and reactive oxygen species play distinct roles in the inflammatory response. Inducible nitric oxide synthase (iNOS) dysregulation is implicated in a variety of chronic and acute diseases and inhibitors of iNOS show noteworthy results in animal models for septic shock, pain, and other conditions, but failed in clinical trials. Conversely, macrophages play a crucial role to regulate peripheral sensitization, cytokines, and chemokines derived from these cells and are potential novel therapeutic targets. Few studies were conducted in order to explore macrophage therapeutics in the animal model and have been shown to prevent/relieve neuropathic pain, but their efficacy has not yet been evaluated in clinical trials. To move evidence-based interventions into practice, pharmacotherapies that target macrophage-driven neuro-inflammation will undoubtedly open up new avenues for beneficial treatment of intractable neuropathic pain.

Concluding this, further research is required to clarify the role of inflammation and the mechanisms that regulate neuropathic pain in fibromyalgia as well as to shed light on potential therapeutic options.

Author details

Vishwas Tripathi[1], Amaresh Mishra[1], Yamini Pathak[1], Aklank Jain[2] and Hridayesh Prakash[3]*

1 School of Biotechnology, Gautam Buddha University, Greater Noida, India

2 Centre for Animal Sciences, School of Basic and Applied Sciences, Central University of Punjab, Bathinda, India

3 Amity Institute of Virology and Immunology, Amity University Uttar Pradesh, Noida, Uttar Pradesh, India

*Address all correspondence to: hprakash@amity.edu

References

[1] Fischer AA. New developments in diagnosis of myofascial pain and fibromyalgia. Physical Medicine and Rehabilitation Clinics of North America. 1997.

[2] Zieglgänsberger W. Substance P and pain chronicity. Cell and Tissue Research. 2019.

[3] Liu Z, Welin M, Bragee B, Nyberg F. A high-recovery extraction procedure for quantitative analysis of substance P and opioid peptides in human cerebrospinal fluid. Peptides. 2000;

[4] Russell IJ, Orr MD, Littman B, Vipraio GA, Alboukrek D, Michalek JE, et al. Elevated cerebrospinal fluid levels of substance p in patients with the fibromyalgia syndrome. Arthritis Rheum. 1994;

[5] Sluka KA, Clauw DJ. Neurobiology of fibromyalgia and chronic widespread pain. Neuroscience. 2016.

[6] Wood PB, Schweinhardt P, Jaeger E, Dagher A, Hakyemez H, Rabiner EA, et al. Fibromyalgia patients show an abnormal dopamine response to pain. Eur J Neurosci. 2007;

[7] Coskun Benlidayi I. Role of inflammation in the pathogenesis and treatment of fibromyalgia. Rheumatology International. 2019.

[8] Lubos E, Handy DE, Loscalzo J. Role of oxidative stress and nitric oxide in atherothrombosis. Frontiers in Bioscience. 2008.

[9] Duque GA, Descoteaux A. Macrophage cytokines: Involvement in immunity and infectious diseases. Frontiers in Immunology. 2014.

[10] Totsch SK, Sorge RE. Immune system involvement in specific pain conditions. Molecular Pain. 2017.

[11] Biswas SK, Mantovani A. Macrophage plasticity and interaction with lymphocyte subsets: Cancer as a paradigm. Nature Immunology. 2010.

[12] Sica A, Mantovani A. Macrophage plasticity and polarization: In vivo veritas. Journal of Clinical Investigation. 2012.

[13] Kiguchi N, Kobayashi Y, Saika F, Sakaguchi H, Maeda T, Kishioka S. Peripheral interleukin-4 ameliorates inflammatory macrophage-dependent neuropathic pain. Pain. 2015;

[14] Kiguchi N, Sakaguchi H, Kadowaki Y, Saika F, Fukazawa Y, Matsuzaki S, et al. Peripheral administration of interleukin-13 reverses inflammatory macrophage and tactile allodynia in mice with partial sciatic nerve ligation. J Pharmacol Sci. 2017;

[15] Komori T, Morikawa Y, Inada T, Hisaoka T, Senba E. Site-specific subtypes of macrophages recruited after peripheral nerve injury. Neuroreport. 2011;

[16] Arout CA, Sofuoglu M, Bastian LA, Rosenheck RA. Gender Differences in the Prevalence of Fibromyalgia and in Concomitant Medical and Psychiatric Disorders: A National Veterans Health Administration Study. J Women's Heal. 2018;

[17] Goldenberg DL, Bradley LA, Arnold LM, Glass JM, Clauw DJ. Understanding fibromyalgia and its related disorders. In: Primary Care Companion to the Journal of Clinical Psychiatry. 2008.

[18] Brill S, Ablin JN, Goor-Aryeh I, Hyat K, Slefer A, Buskila D. Prevalence of fibromyalgia syndrome in patients referred to a tertiary pain clinic. J Investig Med. 2012;

[19] Sarzi-Puttini P, Atzeni F, Mease PJ. Chronic widespread pain: From peripheral to central evolution. Best Practice and Research: Clinical Rheumatology. 2011.

[20] Schmidt-Wilcke T, Clauw DJ. Fibromyalgia: From pathophysiology to therapy. Nature Reviews Rheumatology. 2011.

[21] Celik M, Labuz D, Keye J, Glauben R, Machelska *H. IL*-4 induces M2 macrophages to produce sustained analgesia via opioids. JCI Insight. 2020;

[22] Vanderwall AG, Milligan ED. CytokinesinPain:HarnessingEndogenous Anti-Inflammatory Signaling for Improved Pain Management. Frontiers in Immunology. 2019.

[23] Buskila D, Sarzi-Puttini P. Biology and therapy of fibromyalgia: Genetic aspects of fibromyalgia syndrome. Arthritis Research and Therapy. 2006.

[24] Russell IJ, Larson AA. Neurophysiopathogenesis of Fibromyalgia Syndrome: A Unified Hypothesis. Rheumatic Disease Clinics of North America. 2009.

[25] Geenen R, Jacobs JWG, Bijlsma JWJ. Evaluation and management of endocrine dysfunction in fibromyalgia. Rheumatic Disease Clinics of North America. 2002.

[26] Bote ME, Garca JJ, Hinchado MD, Ortega E. Inflammatory/stress feedback dysregulation in women with fibromyalgia. Neuroimmunomodulation. 2012;

[27] Bote ME, Garcia JJ, Hinchado MD, Ortega E. Fibromyalgia: Anti-Inflammatory and Stress Responses after Acute Moderate Exercise. PLoS One. 2013;

[28] McLean SA, Williams DA, Stein PK, Harris RE, Lyden AK, Whalen G, et al. Cerebrospinal fluid corticotropin-releasing factor concentration is associated with pain but not fatigue symptoms in patients with fibromyalgia. Neuropsychopharmacology. 2006;

[29] Theoharides TC, Stewart JM, Hatziagelaki E. Brain "fog," inflammation a nd obesity: Key aspects of neuropsychiatric disorders improved by luteolin. Front Neurosci. 2015;

[30] Theoharides TC, Petra AI, Stewart JM, Tsilioni I, Panagiotidou S, Akin C. High serum corticotropin-releasing hormone (CRH) and bone marrow mast cell CRH receptor expression in a mastocytosis patient. J Allergy Clin Immunol. 2014;

[31] Giovengo SL, Russell IJ, Larson AA. Increased concentrations of nerve growth factor in cerebrospinal fluid of patients with fibromyalgia. J Rheumatol. 1999;

[32] Lewin GR, Lechner SG, Smith ESJ. Nerve growth factor and nociception: From experimental embryology to new analgesic therapy. Handb Exp Pharmacol. 2014;

[33] Lucas HJ, Brauch CM, Settas L, Theoharides TC. Fibromyalgia - New concepts of pathogenesis and treatment. International Journal of Immunopathology and Pharmacology. 2006.

[34] Pollack S. Mast cells in fibromyalgia. Clinical and Experimental Rheumatology. 2015.

[35] Galli SJ, Tsai M, Piliponsky AM. The development of allergic inflammation. Nature. 2008.

[36] Theoharides TC, Alysandratos KD, Angelidou A, DelivanisDA, SismanopoulosN, ZhangB, et al. Mast cells and inflammation. Biochimica et Biophysica Acta - Molecular Basis of Disease. 2012.

[37] Héron A, Dubayle D. A focus on mast cells and pain. Journal of Neuroimmunology. 2013.

[38] Chatterjea D, Martinov T. Mast cells: Versatile gatekeepers of pain. Molecular Immunology. 2015.

[39] Theoharides TC, Valent P, Akin C. Mast cells, mastocytosis, and related disorders. New England Journal of Medicine. 2015.

[40] Abbadie C. Chemokines, chemokine receptors and pain. Trends in Immunology. 2005.

[41] Charo IF, Ransohoff RM. Mechanisms of disease: The many roles of chemokines and chemokine receptors in inflammation. New England Journal of Medicine. 2006.

[42] Ross RL, Jones KD, Bennett RM, Ward RL, Druker BJ, Wood LJ. Preliminary Evidence of Increased Pain and Elevated Cytokines in Fibromyalgia Patients with Defective Growth Hormone Response to Exercise. Open Immunol J. 2014;

[43] Kadetoff D, Lampa J, Westman M, Andersson M, Kosek E. Evidence of central inflammation in fibromyalgia - Increased cerebrospinal fluid interleukin-8 levels. J Neuroimmunol. 2012;

[44] Rodriguez-Pintó I, Agmon-Levin N, Howard A, Shoenfeld Y. Fibromyalgia and cytokines. Immunol Lett. 2014;

[45] Meng X, Zhang Y, Lao L, Saito R, Li A, Bäckman CM, et al. Spinal interleukin-17 promotes thermal hyperalgesia and NMDA NR1 phosphorylation in an inflammatory pain rat model. Pain. 2013;

[46] Liu Y, Ho RCM, Mak A. The role of interleukin (IL)-17 in anxiety and depression of patients with rheumatoid arthritis. Int J Rheum Dis. 2012;

[47] Gonçalves dos Santos G, Delay L, Yaksh TL, Corr M. Neuraxial Cytokines in Pain States. Frontiers in Immunology. 2020.

[48] von Hehn CA, Baron R, Woolf CJ. Deconstructing the Neuropathic Pain Phenotype to Reveal Neural Mechanisms. Neuron. 2012.

[49] Keller AF, Beggs S, Salter MW, De Koninck Y. Transformation of the output of spinal lamina I neurons after nerve injury and microglia stimulation underlying neuropathic pain. Mol Pain. 2007;

[50] Haroutounian S, Nikolajsen L, Bendtsen TF, Finnerup NB, Kristensen AD, Hasselstrøm JB, et al. Primary afferent input critical for maintaining spontaneous pain in peripheral neuropathy. Pain. 2014;

[51] Calvo M, Dawes JM, Bennett DLH. The role of the immune system in the generation of neuropathic pain. The Lancet Neurology. 2012.

[52] Beggs S, Trang T, Salter MW. P2X4R + microglia drive neuropathic pain. Nature Neuroscience. 2012.

[53] Grace PM, Hutchinson MR, Maier SF, Watkins LR. Pathological pain and the neuroimmune interface. Nature Reviews Immunology. 2014.

[54] Ji RR, Xu ZZ, Gao YJ. Emerging targets in neuroinflammation-driven chronic pain. Nature Reviews Drug Discovery. 2014.

[55] V.A. R, L.P. C, Y. S. Neuroimmunology: What role for autoimmunity, neuroinflammation, and small fiber neuropathy in fibromyalgia, chronic fatigue syndrome, and adverse events after human papillomavirus vaccination? Int J Mol Sci. 2019;

[56] Bäckryd E, Tanum L, Lind AL, Larsson A, Gordh T. Evidence of

both systemic inflammation and neuroinflammation in fibromyalgia patients, as assessed by a multiplex protein panel applied to the cerebrospinal fluid and to plasma. J Pain Res. 2017;

[57] Dong H, Zhang X, Wang Y, Zhou X, Qian Y, Zhang S. Suppression of Brain Mast Cells Degranulation Inhibits Microglial Activation and Central Nervous System Inflammation. Mol Neurobiol. 2017;

[58] Zhang X, Wang Y, Dong H, Xu Y, Zhang S. Induction of microglial activation by mediators released from mast cells. Cell Physiol Biochem. 2016;

[59] Aguzzi A, Barres BA, Bennett ML. Microglia: Scapegoat, saboteur, or something else? Science. 2013.

[60] Réu P, Khosravi A, Bernard S, Mold JE, Salehpour M, Alkass K, et al. The Lifespan and Turnover of Microglia in the Human Brain. Cell Rep. 2017;

[61] Groh J, Martini R. Neuroinflammation as modifier of genetically caused neurological disorders of the central nervous system: Understanding pathogenesis and chances for treatment. GLIA. 2017.

[62] Wang W, Ji P, Riopelle RJ, Dow KE. Functional expression of corticotropin-releasing hormone (CRH) receptor 1 in cultured rat microglia. J Neurochem. 2002;

[63] Thonhoff JR, Simpson EP, Appel SH. Neuroinflammatory mechanisms in amyotrophic lateral sclerosis pathogenesis. Current opinion in neurology. 2018.

[64] Hansson E. Long-term pain, neuroinflammation and glial activation. Scandinavian Journal of Pain. 2010.

[65] Blaszczyk L, Maître M, Lesté-Lasserre T, Clark S, Cota D, Oliet SHR, et al. Sequential alteration of microglia and astrocytes in the rat thalamus following spinal nerve ligation. J Neuroinflammation. 2018;

[66] Blottner D, Huang Y, Trautmann G, Sun L. The fascia: Continuum linking bone and myofascial bag for global and local body movement control on Earth and in Space. A scoping review. REACH. 2019.

[67] Bordoni B, Lintonbon D, Morabito B. Meaning of the Solid and Liquid Fascia to Reconsider the Model of Biotensegrity. Cureus. 2018;

[68] Hannibal KE, Bishop MD. Chronic stress, cortisol dysfunction, and pain: A psychoneuroendocrine rationale for stress management in pain rehabilitation. Phys Ther. 2014;

[69] Neyal M, Yimenicioglu F, Aydeniz A, Taskin A, Saglam S, Cekmen M, et al. Plasma nitrite levels, total antioxidant status, total oxidant status, and oxidative stress index in patients with tension-type headache and fibromyalgia. Clin Neurol Neurosurg. 2013;

[70] Bozkurt M, Caglayan M, Oktayoglu P, Em S, Batmaz I, Sariyildiz MA, et al. Serum prolidase enzyme activity and oxidative status in patients with fibromyalgia. Redox Rep. 2014;

[71] Ashina M, Bendtsen L, Jensen R, Lassen LH, Sakai F, Olesen J. Possible mechanisms of action of nitric oxide synthase inhibitors in chronic tension-type headache. Brain. 1999;

[72] Ghafourifar P, Cadenas E. Mitochondrial nitric oxide synthase. Trends in Pharmacological Sciences. 2005.

[73] Ozgocmen S, Ozyurt H, Sogut S, Akyol O, Ardicoglu O, Yildizhan H. Antioxidant status, lipid peroxidation

and nitric oxide in fibromyalgia: Etiologic and therapeutic concerns. Rheumatol Int. 2006;

[74] Kone BC, Kuncewicz T, Zhang W, Yu ZY. Protein interactions with nitric oxide synthases: Controlling the right time, the right place, and the right amount of nitric oxide. American Journal of Physiology - Renal Physiology. 2003.

[75] Sharma JN, Al-Omran A, Parvathy SS. Role of nitric oxide in inflammatory diseases. Inflammopharmacology. 2007.

[76] Levy D, Höke A, Zochodne DW. Local expression of inducible nitric oxide synthase in an animal model of neuropathic pain. Neurosci Lett. 1999;

[77] Kawahara K, Hohjoh H, Inazumi T, Tsuchiya S, Sugimoto Y. Prostaglandin E2-induced inflammation: Relevance of prostaglandin e receptors. Biochimica et Biophysica Acta - Molecular and Cell Biology of Lipids. 2015.

[78] Santa-Cecília F V., Ferreira DW, Guimaraes RM, Cecilio NT, Fonseca MM, Lopes AH, et al. The NOD2 signaling in peripheral macrophages contributes to neuropathic pain development. Pain. 2019;

[79] Thacker MA, Clark AK, Marchand F, McMahon SB. Pathophysiologyofperipheralneuropathic pain: Immune cells and molecules. Anesthesia and Analgesia. 2007.

[80] Kiguchi N, Kobayashi Y, Kishioka S. Chemokines and cytokines in neuroinflammation leading to neuropathic pain. Current Opinion in Pharmacology. 2012.

[81] Ristoiu V. Contribution of macrophages to peripheral neuropathic pain pathogenesis. Life Sciences. 2013.

[82] Okifuji A, Gao J, Bokat C, Hare BD. Management of fibromyalgia syndrome in 2016. Pain management. 2016.

[83] Häuser W, Bernardy K, Arnold B, Offenbächer M, Schiltenwolf M. Efficacy of multicomponent treatment in fibromyalgia syndrome: A meta-analysis of randomized controlled clinical trials. Arthritis Care Res. 2009;

[84] Üçeyler N, Valenza R, Stock M, Schedel R, Sprotte G, Sommer C. Reduced levels of antiinflammatory cytokines in patients with chronic widespread pain. Arthritis Rheum. 2006;

[85] Sturgill J, McGee E, Menzies V. Unique cytokine signature in the plasma of patients with fibromyalgia. J Immunol Res. 2014;

[86] Clauw DJ. Fibromyalgia: A clinical review. JAMA - Journal of the American Medical Association. 2014.

[87] Goldenberg DL, Burckhardt C, Crofford L. Management of fibromyalgia syndrome. Journal of the American Medical Association. 2004.

[88] Fitzcharles MA, Ste-Marie PA, Goldenberg DL, Pereira JX, Abbey S, Choinière M, et al. Canadian pain society and canadian rheumatology association recommendations for rational care of persons with fibromyalgia. A summary report. J Rheumatol. 2013;

[89] Busch AJ, Webber SC, Richards RS, Bidonde J, Schachter CL, Schafer LA, et al. Resistance exercise training for fibromyalgia. Cochrane Database of Systematic Reviews. 2013.

[90] Bidonde J, Busch AJ, Webber SC, Schachter CL, Danyliw A, Overend TJ, et al. Aquatic exercise training for fibromyalgia. Cochrane Database of Systematic Reviews. 2014.

[91] Mist SD, Firestone KA, Jones KD. Complementary and alternative exercise for fibromyalgia: A meta-analysis. Journal of Pain Research. 2013.

[92] Sawynok J, Lynch M. Qigong and fibromyalgia: Randomized controlled trials and beyond. Evidence-based Complementary and Alternative Medicine. 2014.

[93] Porter NS, Jason LA, Boulton A, Bothne N, Coleman B. Alternative medical interventions used in the treatment and management of myalgic encephalomyelitis/chronic fatigue syndrome and fibromyalgia. J Altern Complement Med. 2010;

[94] Arranz LI, Canela MÁ, Rafecas M. Dietary aspects in fibromyalgia patients: Results of a survey on food awareness, allergies, and nutritional supplementation. Rheumatol Int. 2012;

[95] Cordero MD, Cano-García FJ, Alcocer-Gómez E, de Miguel M, Sánchez-Alcázar JA. Oxidative stress correlates with headache symptoms in Fibromyalgia: Coenzyme Q 10 effect on clinical improvement. PLoS One. 2012;

[96] Cordero MD, Cotán D, del-Pozo-Martín Y, Carrión AM, de Miguel M, Bullón P, et al. Oral coenzyme Q10 supplementation improves clinical symptoms and recovers pathologic alterations in blood mononuclear cells in a fibromyalgia patient. Nutrition. 2012;

[97] Rossini M, Di Munno O, Valentini G, Bianchi G, Biasi G, Cacace E, et al. Double-blind, multicenter trial comparing acetyl l-carnitine with placebo in the treatment of fibromyalgia patients. Clin Exp Rheumatol. 2007;

[98] Middleton E, Kandaswami C, Theoharides TC. The effects of plant flavonoids on mammalian cells: Implications for inflammation, heart disease, and cancer. Pharmacological Reviews. 2000.

[99] Cazarolli L, Zanatta L, Alberton E, Bonorino Figueiredo MS, Folador P, Damazio R, et al. Flavonoids: Prospective Drug Candidates. Mini-Reviews Med Chem. 2008;

[100] Izzi V, Masuelli L, Tresoldi I, Sacchetti P, Modesti A, Galvano F, et al. The effects of dietary flavonoids on the regulation of redox inflammatory networks. Frontiers in Bioscience. 2012.

[101] Kimata M, Shichijo M, Miura T, Serizawa I, Inagaki N, Nagai H. Effects of luteolin, quercetin and baicalein on immunoglobulin E- mediated mediator release from human cultured mast cells. Clin Exp Allergy. 2000;

[102] Jäger AK, Saaby L. Flavonoids and the CNS. Molecules. 2011.

[103] Grosso C, Valentão P, Ferreres F, Andrade P. The Use of Flavonoids in Central Nervous System Disorders. Curr Med Chem. 2013;

[104] Macfarlane GJ, Kronisch C, Dean LE, Atzeni F, Häuser W, Flub E, et al. EULAR revised recommendations for the management of fibromyalgia. Ann Rheum Dis. 2017;

[105] Häuser W, Arnold B, Eich W, Felde E, Flügge C, Henningsen P, et al. Management of fibromyalgia syndrome--an interdisciplinary evidence-based guideline. Ger Med Sci. 2008;

[106] Häuser W, Petzke F, Üeyler N, Sommer C. Comparative efficacy and acceptability of amitriptyline, duloxetine and milnacipran in fibromyalgia syndrome: A systematic review with meta-analysis. Rheumatology. 2011;

[107] Clemons A, Vasiadi M, Kempuraj D, Kourelis T, Vandoros G, Theoharides TC. Amitriptyline and prochlorperazine inhibit proinflammatory mediator release from human mast cells: Possible relevance to chronic fatigue syndrome. Journal of Clinical Psychopharmacology. 2011.

[108] Harris RE, Napadow V, Huggins JP, Pauer L, Kim J, Hampson J, et al. Pregabalin rectifies aberrant brain chemistry, connectivity, and functional response in chronic pain patients. Anesthesiology. 2013;

[109] Häuser W, Bernardy K, Üçeyler N, Sommer C. Treatment of fibromyalgia syndrome with gabapentin and pregabalin - A meta-analysis of randomized controlled trials. Pain. 2009;

[110] Häuser W, Bernardy K, Üçeyler N, Sommer C. Treatment of fibromyalgia syndrome with antidepressants: A meta-analysis. JAMA - Journal of the American Medical Association. 2009.

[111] Moore RA, Wiffen PJ, Derry S, Rice ASC. Gabapentin for chronic neuropathic pain and fibromyalgia in adults. Cochrane Database of Systematic Reviews. 2014.

[112] Derry S, Wiffen PJ, Häuser W, Mücke M, Tölle TR, Bell RF, et al. Oral nonsteroidal anti-inflammatory drugs for fibromyalgia in adults. Cochrane Database of Systematic Reviews. 2017.

[113] Moldofsky H, Harris HW, Tad Archambault W, Kwong T, Lederman S. Effects of bedtime very low dose cyclobenzaprine on symptoms and sleep physiology in patients with fibromyalgia syndrome: A double-blind randomized placebo-controlled study. J Rheumatol. 2011;

[114] Lunn MPT, Hughes RAC, Wiffen PJ. Duloxetine for treating painful neuropathy, chronic pain or fibromyalgia. Cochrane Database of Systematic Reviews. 2014.

[115] Arnold LM, Rosen A, Pritchett YL, D'Souza DN, Goldstein DJ, Iyengar S, et al. A randomized, double-blind, placebo-controlled trial of duloxetine in the treatment of women with fibromyalgia with or without major depressive disorder. Pain. 2005;

[116] Russell IJ. Fibromyalgia syndrome: Approach to management. CNS Spectrums. 2008.

[117] Calandre EP, Rico-Villademoros F, Slim M. An update on pharmacotherapy for the treatment of fibromyalgia. Expert Opinion on Pharmacotherapy. 2015.

[118] Welsch P, Üçeyler N, Klose P, Walitt B, Häuser W. Serotonin and noradrenaline reuptake inhibitors (SNRIs) for fibromyalgia. Cochrane Database of Systematic Reviews. 2018.

[119] Krell H V., Leuchter AF, Cook IA, Abrams M. Evaluation of reboxetine, a noradrenergic antidepressant, for the treatment of fibromyalgia and chronic low back pain. Psychosomatics. 2005.

[120] Arnold LM, Hirsch I, Sanders P, Ellis A, Hughes B. Safety and efficacy of esreboxetine in patients with fibromyalgia: A fourteen-week, randomized, double-blind, placebo-controlled, multicenter clinical trial. Arthritis Rheum. 2012;

[121] Ablin J, Fitzcharles MA, Buskila D, Shir Y, Sommer C, Häuser W. Treatment of fibromyalgia syndrome: Recommendations of recent evidence-based interdisciplinary guidelines with special emphasis on complementary and alternative therapies. Evidence-based Complementary and Alternative Medicine. 2013.

[122] Crofford LJ, Rowbotham MC, Mease PJ, Russell IJ, Dworkin RH,

Corbin AE, et al. Pregabalin for the treatment of Fibromyalgia syndrome: Results of a randomized, double-blind, placebo-controlled trial. Arthritis Rheum. 2005;

[123] Ohta H, Oka H, Usui C, Ohkura M, Suzuki M, Nishioka K. A randomized, double-blind, multicenter, placebo-controlled phase III trial to evaluate the efficacy and safety of pregabalin in Japanese patients with fibromyalgia. Arthritis Res Ther. 2012;

[124] Hearn L, Derry S, Moore RA. Lacosamide for neuropathic pain and fibromyalgia in adults. Cochrane Database Syst Rev. 2012;

[125] Skrabek RQ, Galimova L, Ethans K, Perry D. Nabilone for the Treatment of Pain in Fibromyalgia. J Pain. 2008;

[126] Ware MA, Fitzcharles MA, Joseph L, Shir Y. The effects of nabilone on sleep in fibromyalgia: Results of a randomized controlled trial. Anesth Analg. 2010;

[127] Walitt B, Klose P, Fitzcharles MA, Phillips T, Häuser W. Cannabinoids for fibromyalgia. Cochrane Database of Systematic Reviews. 2016.

[128] Konrad C, Weber J, Schley M, Casutt M, Gerber H, Schuepfer G, et al. Tetrahydrocannabinol (Delta 9-THC) treatment in chronic central neuropathic pain and fibromyalgia patients: Results of a multicenter survey. Anesthesiology Research and Practice. 2009.

[129] Cohen SP, Verdolin MH, Chang AS, Kurihara C, Morlando BJ, Mao J. The Intravenous Ketamine Test Predicts Subsequent Response to an Oral Dextromethorphan Treatment Regimen in Fibromyalgia Patients. J Pain. 2006;

[130] Johnson JW, Kotermanski SE. Mechanism of action of memantine. Current Opinion in Pharmacology. 2006.

[131] Harris RE, Sundgren PC, Craig AD, Kirshenbaum E, Sen A, Napadow V, et al. Elevated insular glutamate in fibromyalgia is associated with experimental pain. Arthritis Rheum. 2009;

[132] Olivan-Blázquez B, Herrera-Mercadal P, Puebla-Guedea M, Pérez-Yus MC, Andrés E, Fayed N, et al. Efficacy of memantine in the treatment of fibromyalgia: A double-blind, randomised, controlled trial with 6-month follow-up. Pain. 2014;

[133] Akinkunle Fadare SO. Relief and Resolution of Fibromyalgia Symptoms with Low Dose Methotrexate – The Origin of Pain is Inflammation and the Inflammatory Response. Rheumatol Curr Res. 2014;

[134] Chakr RM da S, Brenol C, Ranzolin A, Bernardes A, Dalosto AP, Ferrari G, et al. Rheumatoid arthritis seems to have DMARD treatment decision influenced by fibromyalgia. Rev Bras Reumatol (English Ed. 2017;

[135] Salaffi F, Gerardi MC, Atzeni F, Batticciotto A, Talotta R, Draghessi A, et al. The influence of fibromyalgia on achieving remission in patients with long-standing rheumatoid arthritis. Rheumatol Int. 2017;

[136] Wang H, Buchner M, Moser MT, Daniel V, Schiltenwolf M. The role of IL-8 in patients with fibromyalgia: A prospective longitudinal study of 6 months. Clin J Pain. 2009;

[137] Cordero MD, Alcocer-Gómez E, Culic O, Carrión AM, De Miguel M, Díaz-Parrado E, et al. NLRP3 inflammasome is activated in fibromyalgia: The effect of coenzyme Q10. Antioxidants and Redox Signaling. 2014.

[138] Tsai RY, Jang FL, Tai YH, Lin SL, Shen CH, Wong CS. Ultra-low-dose

naloxone restores the antinociceptive effect of morphine and suppresses spinal neuroinflammation in PTX-treated rats. Neuropsychopharmacology. 2008;

[139] Liu B, Jiang JW, Wilson BC, Du L, Yang SN, Wang JY, et al. Systemic infusion of naloxone reduces degeneration of rat substantia nigral dopaminergic neurons induced by intranigral injection of lipopolysaccharide. J Pharmacol Exp Ther. 2000;

[140] Greeneltch KM, Haudenschild CC, Keegan AD, Shi Y. The opioid antagonist naltrexone blocks acute endotoxic shock by inhibiting tumor necrosis factor-α production. Brain Behav Immun. 2004;

[141] Younger J, Mackey S. Fibromyalgia symptoms are reduced by low-dose naltrexone: A pilot study. Pain Med. 2009;

[142] Wang H, Yu M, Ochani M, Amelia CA, Tanovic M, Susarla S, et al. Nicotinic acetylcholine receptor α7 subunit is an essential regulator of inflammation. Nature. 2003;

[143] de Jonge WJ, van der Zanden EP, The FO, Bijlsma MF, van Westerloo DJ, Bennink RJ, et al. Stimulation of the vagus nerve attenuates macrophage activation by activating the Jak2-STAT3 signaling pathway. Nat Immunol. 2005;

[144] Tracey KJ. Physiology and immunology of the cholinergic antiinflammatory pathway. Journal of Clinical Investigation. 2007.

[145] Borovikova L V., Ivanova S, Zhang M, Yang H, Botchkina GI, Watkins LR, et al. Vagus nerve stimulation attenuates the systemic inflammatory response to endotoxin. Nature. 2000;

[146] Ulloa L. The vagus nerve and the nicotinic anti-inflammatory pathway. Nature Reviews Drug Discovery. 2005.

[147] Wang H, Liao H, Ochani M, Justiniani M, Lin X, Yang L, et al. Cholinergic agonists inhibit HMGB1 release and improve survival in experimental sepsis. Nat Med. 2004;

[148] Saeed RW, Varma S, Peng-Nemeroff T, Sherry B, Balakhaneh D, Huston J, et al. Cholinergic stimulation blocks endothelial cell activation and leukocyte recruitment during inflammation. J Exp Med. 2005;

Section 4

Macrophages in Stroke

Chapter 6

Microglial Plasticity Contributes to Recovery of Bone Marrow Mononuclear Cells during Experimental Stroke

Edna Cristina S. Franco, Marcelo Marques Cardoso, Celice Cordeiro de Souza, Michelle Castro da Silva, Carolina Ramos dos Santos and Walace Gomes-Leal

Abstract

Brain stroke is an acute neural disorder characterized by obstruction (ischemic) or rupture (hemorrhagic) of blood vessels causing neural damage and subsequent functional impairment. Its pathophysiology is complex and involves a multitude of pathological events including energetic collapse, excitotoxicity, oxidative stress, metabolic acidosis, cell death and neuroinflammation. Despite its clinical importance, there is no effective pharmacological therapies available to diminish secondary damage avowing functional deficits. Considering the failure of pharmacological approaches for stroke, cell therapy came as promising alternative. Different cell types have been investigated in different experimental models with promising results. An important issue regarding the transplantation of stem cells into the damaged CNS tissue is how the pathological environment influences the transplanted cells. It has been established that an exacerbated inflammation in the pathological environment is detrimental to the survival of the transplanted stem cells. This prompted us to develop an experimental strategy to improve the therapeutic actions of bone marrow mononuclear cells (BMMCs) transplanted into the acute phase of brain stroke by modulating microglial activation with minocycline. In this chapter, we first review the basic pathophysiology of ischemic stroke with emphasis on the role of microglia to the pathological outcome. We then review the experimental approach of modulating microglia activation in order to enhance therapeutic actions of BMMCS for experimental stroke. We suggest that such an approach may be applied as an adjuvant therapy to control excessive neuroinflammation in the pathological environment allowing acute transplants and improving therapeutic actions of different kind of stem cells.

Keywords: stroke, stem cells, cell therapy, minocycline, neuroinflammation, neuroprotection

1. Introduction

The central nervous system (CNS) is affected by acute and chronic neural disorders. In acute neural disorders, like stroke, spinal cord injury (SCI) and brain trauma, neuronal and glial loss happens quickly with inexorable cell loss and functional impairment [1–5]. In chronic neurodegenerative diseases, including Parkinson's, Huntington's, Alzheimer's diseases and Amyotrophic Lateral Sclerosis (ALS) progressive cell loss occurs over decades with inexorable functional loss and sensory-motor and/or cognitive declines [1, 5].

Stroke is an acute neural disorder and leading cause of death and functional impairment worldwide [1, 5]. Recent epidemiological data point out that occurred 1.12 million cases of stroke in 2017 in European Union countries, with 9.53 million survivors, approximately half a million deaths, and 7 million people with permanent sequelae [6]. According to this study there will be about 40,000 new stroke cases in Europe by 2047, and an increase of about 27% in the number of people living with sequelae of some type of stroke [6].

Similar data were published by the American Heart Association (AHA), which showed that about 7 million Americans over the age of 20 had strokes between 2013 and 2016 with a prevalence that increases with advancing age in both sexes [7]. The same study shows that more than 3.4 million Americans over the age of 20 will have a stroke by 2030, an increase of 20.5% in prevalence compared to 2012.

Stroke is a vascular disorder characterized by obstruction (ischemic) or rupture of blood vessels (hemorrhagic). Following this primary pathological event, further outcomes are diverse and characterized by a multitude of factors, excitotoxicity, oxidative stress, metabolic acidosis, periinfarct depolarization, apoptosis and uncontrolled neuroinflammation, which contributes to cell death and functional impairment in both experimental animals and humans [1–5, 8–10].

There are no effective pharmacological treatment or cell therapy approved for stroke [2, 5, 8]. Approved clinical therapy is restricted to thrombolysis by using the recombinant tissue plasminogen activator (tPA) for ischemic stroke, which is limited by its narrow therapeutic window [11–13]. In the clinical practice, people with stroke arrive at the hospital usually several hours after the onset of symptoms, outside the therapeutic window for the use of thrombolytic agents (alteplase), mainly in low income countries with a limited public health system.

Numerous experimental studies have shown the inefficacy of several tested neuroprotective agents, including glutamatergic antagonists, calcium antagonists, antioxidants, magnesium for inducing neuroprotection in animals [14, 15]. This fact raised considerable skepticism regarding the possibility of finding an effective neuroprotective agent for neurological human diseases [14, 15].

Considering the limitations of pharmacological approaches, it is believed that cell therapy is considered a promising therapeutic approach for inducing neuroprotection, cell replacement and functional improvement following both acute and chronic neural disorders [16–20]. This is confirmed by several studies using experimental models of neural disorders, including stroke [21].

Different types of stem cells from different sources (umbilical cord blood cells, bone marrow stem cells, neural stem cells, induced pluripotent stem cells) have been tested in different experimental stroke models rendering neuroprotection and functional impairment [16–20].

Although embryonic stem cell transplantation is considered a promising future therapeutic approach for neural disorders, technical and ethical-legal restrictions have hindered its clinical use [16–20]. Stem/progenitor cells derived from adult sources, including bone marrow derived stem cells (BMSCs), have been transplanted in both acute and subacute phase after stroke affording considerable degree of neuroprotection [22–26].

An important issue regarding the transplantation of stem cells into the damaged CNS tissue is how the pathological environment influences the transplanted cells. In disorders like stroke and trauma, an intense inflammatory response is elicited with both cellular and humoral components belonging to innate and adaptive immune systems.

It has been previously shown that bone marrow mononuclear cells (BMMCs) transplanted into the intact adult rodent brain are rejected by components of the CNS inflammatory response [27]. In addition, it has been shown that brain macrophages impair survival and integration of embryonic stem cells transplanted into the acute phase of brain trauma [28]. This prompted us to develop an experimental strategy to improve the therapeutic actions of BMMCs transplanted into the acute phase of brain stroke by modulating microglial activation with minocycline [22, 23]. Using this approach, we were successful in improving therapeutic actions of BMMC transplanted into both ischemic cortex [22] and striatum {Cardoso, 2013 #27 in adult rats.

In this chapter, we will first review the basic pathophysiology of ischemic stroke with emphasis on the role of microglia to the pathological outcome. We then review the experimental approach of modulating microglia activation in order to enhance therapeutic actions of BMMCS for experimental stroke. We suggest that such an approach may be applied as an adjuvant therapy to control excessive neuroinflammation in the pathological environment allowing acute transplants and improving therapeutic actions of different kind of stem cells.

2. Stroke pathophysiology

2.1 Overview

The pathophysiological events of stroke are extremely complex and involve different mechanisms [1–5, 8]. Following metabolic collapse in the brain function, ischemic injury results in a complex sequence of pathophysiological events that include metabolic acidosis, excitotoxicity, peri-infarction depolarization, oxidative stress, programmed cell death and neuroinflammation [1–5, 8].

Several events are related to cell death after stroke. The interruption of blood flow generates an energy collapse in the cells, followed by ionic imbalance, with intense Ca^{2+} influx, exacerbated release of glutamate and oxidative/nitrosative stress. All of these events are correlated and lead to cell death, triggering an intense inflammatory response in the ischemic environment, which has a dubious role, as it can contribute to both tissue repair and to intensify the injury [9, 10, 29].

The brain tissue requires a high energy demand for its optimal functioning, being responsible for 20% of all the body's oxygen consumption. In addition to the energy needed to maintain cellular homeostasis, the synaptic transmission process requires a large amount of ATP, which is obtained from the oxidation of glucose in the mitochondria oxidative phosphorylation chain. Therefore, the glucose and oxygen reduction in the ischemic environment has severe deleterious effects on the nervous tissue [30].

After ischemia, the alteration of several physiological, biochemical, molecular and genetic mechanisms results in cell death and impaired neuronal function (**Figure 1**). In the ischemic core, cell death occurs predominantly from necrosis minutes after ischemia. Initially, the interruption of blood supply leads to a reduction in oxygen and glucose reaching neurons, which compromises the process of oxidative phosphorylation in mitochondria and drastically reduces the ATP production [1, 5, 30].

This reduction impairs the functioning of ATP-dependent ion pumps, such as the Na+/K+ pump, causing an imbalance in the ionic potential and generating the cell membrane depolarization. In addition, impaired mitochondrial function

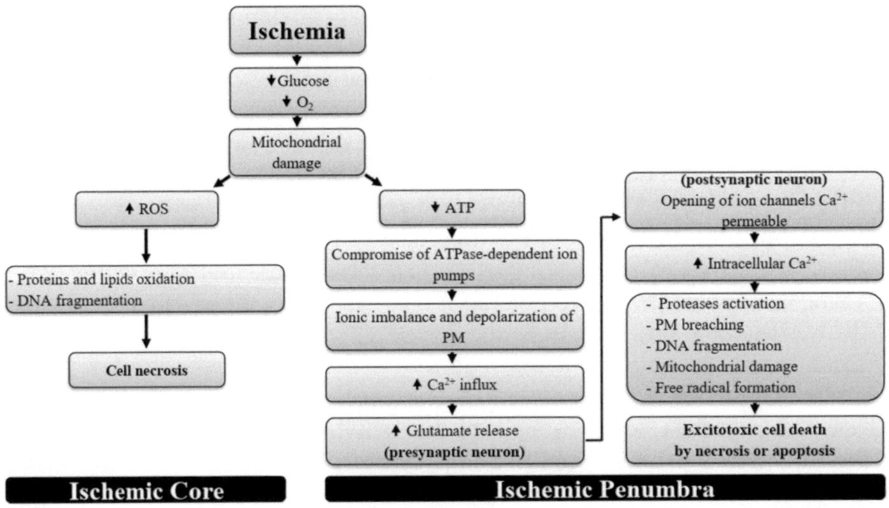

Figure 1.
Overview of stroke pathophysiology. The primary pathological event is abrupt reduction of blood flow resulting in oxygen level shortage, mithochondrial damage and ATP depletion. Metabolical colapse induce formation of reactive oxygen species (ROS) leading to protein and lipid oxidation and cell death. Control f excitatory neurotransmitter levels is lost leading to increased calcium intracelular levels, pathological activation of lipases, proteases and further cell death.

generates the production of superoxide radicals, reducing antioxidant activity and causing oxidative stress, which in turn results in the oxidation of proteins and lipids in the cell membrane and DNA fragmentation, ultimately leading to cell necrosis [30]. On the other hand, the events that lead to cell death in the penumbra area are more complex and can extend for weeks after ischemia [30–32].

In general, the oxygen and glucose reduction generates an imbalance in the ionic potential of the membrane, which causes the intense influx of Ca2+ and the release of glutamate, events that are interspersed in a positive feedback loop and lead to cell death due to excitotoxicity. Since, in this pathological environment in which the mechanisms of intracellular Ca2+ control are compromised, cell death programmed by apoptosis and/or total cell collapse may occur, leading to necrosis [33]. The main events triggered by the scarcity of glucose and oxygen that lead to cell death in the center and in the ischemic penumbra are described in **Figure 1**.

2.2 Cell death after stroke

Several mechanisms of cell death can be triggered after ischemia, including necrosis and apoptosis, which can occur interchangeably according to changes in the ischemic environment. Necrosis, predominant in the ischemic core, is characterize by the cytoplasm vacuolization, cell edema, plasma membrane rupture and pro-inflammatory cytokines release [30–33]. Apoptosis is strictly regulated, demands energy, being predominant in the ischemic penumbra area, and it is characterized by cell retraction, chromatin density and condensation increased nuclear membrane rupture and formation of the apoptotic bodies, but maintaining the membrane cellular integrity [33, 34].

Apoptosis occurs intrinsically, by mitochondrial signaling, or extrinsically, by cell death receptors stimulating, such as tumor necrosis factor α (TNF-α), TRAIL receptors (TNF-related apoptosis-inducing ligand) and FAS (CD95/APO1). In both processes, it is necessary to activate the cysteine-aspartate protease family proteins, called caspases [35]. This activation involves the Bcl-2 family proteins, which includes pro-apoptotic proteins (Bax and Bak) and anti-apoptotic proteins (Bid and

Bcl-2) acting at external mitochondrial membrane maintenance and Ca2+ regulation in the mitochondria and endoplasmic reticulum [33]. After stroke, neuronal death from apoptosis occurs primarily intrinsically due to mitochondrial damage and cytochrome c release in the cytosol [33, 35].

Another process present in stroke is the autophagy, a programmed cell death in which cell degradation is carried out by lysosomes in response to severe cell damage when the cell is submitted to environmental stress. This can occur in three ways: mediated by chaperones, microautophagy and macroautophagy, the most observed in stroke. In general, autophagy is blocked by activated mTOR (target of rapamycin) and induced by AMP-activated protein kinase (AMPK) and rapamycin that inhibits mTOR. The moderate autophagy activation is a beneficial and anti-apoptotic process, including the mitochondria removal from damaged cells. On the other hand, this process becomes deleterious and pro-apoptotic when intensified in the ischemic environment, being related to the inflammatory process [33].

Two mechanisms of cell death present in stroke are not fully described: necroptosis and pyroptosis. Necroptosis has characteristics similar to necrosis, such as cell edema, plasma membrane rupture and pro-inflammatory cytokines release, however, it is not a completely passive process and it is activated through the receptors of cell death, such as TNF-α, and inhibited by the necroptosis-inhibiting factor-1 (Nec-1) [35]. Pyroptosis is trigger by caspase-1, being characterized by DNA damage, plasma membrane rupture and pro-inflammatory factors release [36].

The different mechanisms of cell death that occur after stroke are correlate in a complex process. Although there is a predominance of certain types in the ischemic core and others in the ischemic penumbra, some pathways occur simultaneously in both area, playing a beneficial, harmful or dubious role. Thus, the results of the interaction between these mechanisms is directly relate to the inflammatory process after ischemia and will define the affected cells survival [33].

2.3 Neuroinflammation

Neuroinflammation is an important component of the pathophysiology of acute and chronic neural disorders [9, 10, 22, 37]. After stroke and trauma, an intense inflammatory response is initiated with both humoral and cellular components [9, 10, 23, 38–42].

The cellular components of neuroinflammation belong to both innate and adaptive immune systems [9, 10]. In experimental models of stroke [9, 10, 38, 41, 42] and trauma [40, 43], neutrophils are recruited from blood vessels to the lesion site, peaking at 24 h post-damage onset. In latter survival times, macrophages dominate the pathological environment peaking between 3 and 7 days after trauma [40, 43] or ischemia [23, 38, 41, 42] in adult rodents.

Macrophages are derived from both resident microglia and blood monocytes recruited from the blood stream [44, 45]. An intense microglial activation is observed in the first week after spinal cord trauma [40, 43] and experimental stroke in both cortex [22], striatum [9, 10, 23, 41, 42].

Microgliosis is accompanied by intense astrocytosis that differs in its temporal profile in different compartments of the CNS [39]. In our previous studies, we demonstrated that astrocytes are activated more quickly in the white matter (WM) than in gray matter (GM) after excitotoxic injury to the spinal cord [39].

The inflammatory response in the CNS has a dubious nature, contributing to events of tissue repair and regeneration, as well as contributing to the exacerbation of the injury process [9, 10]. This is most evident when considering the role of microglial/macrophage cells. It has been shown that microglia inhibition with minocycline induces neuroprotection, decreases axonal loss and programmed cell death after traumatic injury [46–48] or ischemia [23, 49–51].

Considerable neuroprotection is obtained after blocking recruitment of hematogenous macrophages after experimental spinal cord trauma [52]. Some studies suggest that treatment with minocycline is safe and can benefit people in the acute phase of ischemic stroke [53–56]. This fact is even more relevant considering that minocycline, despite having pleiotropic effects, has an important action on microglial activity [23, 49–51] and that, in humans, microglial activation is an important component of neuroinflammatory events [57–59].

Despite the above data, it is known that microglial/macrophage cells can induce neuroprotection after trauma [60–63] or ischemia [64–67]. Recently, we suggested that this dubious action is influenced by the pathological environment and that ligands can activate different receptors in the microglial membrane, activating their harmful and/or protective actions [9, 10].

3. The dual role of microglia after stroke

Neuroinflammation is one of the main components of the pathophysiology of CNS diseases [9, 10, 68–70]. After the stroke, a complex range of humoral and cellular responses occurs, with different consequences for the neuropathological development [9, 10, 68–70]. Neutrophils, lymphocytes and macrophages are recruited to the lesion site, in addition to the concomitant activation of microglia and astrocytes [9, 10]. Concomitantly, an intricate network of humoral responses is developed, characterized by the release of several pro and anti-inflammatory cytokines, with specific roles, depending on the moment after the injury [71].

Neuroinflammation has beneficial and harmful effects after stroke and other neural disorders [9, 10]. The main component of the inflammatory response that occurs after acute neural disorders are the microglia cells, the macrophages residing in the CNS, myeloid cells derived from progenitors of the Yolk sac embryonic structure [72, 73].

Microglial cells are components of the innate immune system that patrol the CNS in normal situations using stochastic movements of its thin and long branches in order to protect it from harmful events [74, 75], movements that depend on endogenous ATP [76].

During development, these cells phagocytose in excess synapses, contributing to the maturation of neural circuits, an action that depends on interleukin 33 released by astrocytes [77]. Like cells of the innate immune system, microglial cells are the first line of defense of the CNS against viruses, bacteria and other pathogens, removing them during phagocytosis infection or by releasing powerful pro-inflammatory agents, nitric oxide, proteases, free radicals in addition to other lytic agents [78–80].

It is well established that after stroke, trauma and other diseases of the CNS, microglial cells have a dubious action, contributing to exacerbation of the injury and repair [9, 10]. The inhibition of microglial activation with tetracycline minocycline decreases the infarction area neuroinflammation, both in the cortex and in the striatum, after experimental occlusion of the middle cerebral artery (MCAO) [51]. Modulation of microglial activation improves the therapeutic effects of bone marrow mononuclear cells, transplanted intravenously after focal ischemic lesion in the cortex [23] or striatum [23]. Paradoxically, the presence of microglial cells of the BV2 cortical lineage in organotypic culture reduces neuron death after glucose and oxygen deprivation [65]. In this same experimental model, microglial cells are highly beneficial for phagocytosing polymorphonuclear cells [66]. After MCAO in mice, ablation of microglial proliferation worsens the inflammatory process and induces higher levels of programmed cell death after focal ischemia [67].

We have proposed that "friendly fire hypothesis" to explain the dual role of microglia after stroke and other neural disorders [9–10]. According to the friendly fire hypothesis microglia used their biochemical machinery normally used to fight infections during sterile inflammation in neural disorders like stroke, trauma and even over the course of chronic neurodegenerative diseases [9, 10]. According to this notion danger signals released by stressed, damaged or dying cells might bind the same pattern recognition receptors (PRRs), or even different receptors, normally activated by pathogen-associated molecular patterns (PAMPs) present in the microglia cell membrane culminating in secondary cell damage [9, 10]. This is supported by our preliminary findings showing that in the presence of bacterial infection ischemic damage is larger than in the absence of infection [9].

4. Stem cell therapy for stroke

Currently, there are no effective pharmacological treatments for stroke [14, 15]. Conventional therapy is restricted to thrombolysis by using the recombinant tissue plasminogen activator (tPA) [11, 13]. Few patients with ischemic stroke are benefited from thrombolytic therapy, mainly because of its narrow therapeutic window [11, 13]. In clinical practice, people with stroke arrive at the hospital, usually several hours after the onset of symptoms, outside the therapeutic window for the use of thrombolytic agents (alteplase), mainly in low-income countries with deficient the health systems.

Numerous experimental studies have shown promising results of experimental drugs, including glutamatergic antagonists, calcium antagonists, antioxidants, magnesium and many others as neuroprotective agents in experimental animals [14]. However, despite the experimental success of these drugs, their application as neuroprotective agents in humans has been totally ineffective [14, 15]. This fact gave rise to great skepticism regarding the possibility of finding an effective neuroprotective agent for neurological diseases in humans [2, 81].

Considering the failure of translational research for achieving an effective neuroprotective agent, cell therapy came up as promising approach for inducing neuroprotection, cell replacement and functional improvement after acute and chronic neural disorders, depending on cell type [8, 17, 21, 26].

Different types of stem cells from different sources (umbilical cord blood cells, bone marrow stem cells, mesenchymal stem cells, neural stem cells, immortalized cell lines, induced pluripotent stem cells (IPSCs) were used in experimental stroke models to afford neuroprotection, cell replacement and subsequent functional recovery [8, 17, 21, 26]. Although embryonic stem cell transplantation is considered a promising treatment for neurodegenerative diseases, technical and ethical-legal restrictions have hindered its clinical use [8, 17, 21, 26].

An alternative source of cell therapy involves the use of adult progenitor cells derived from bone marrow, including mesenchymal stem cells or their faction – the bone marrow mononuclear cells [22–25, 82]. Both mesenchymal and mononuclear bone marrow cells (BMMCs) are highly anti-inflammatory and neuroprotective in experimental models of stroke [22–25, 82, 83].

5. Bone marrow mononuclear cells and stroke

BMMCs are adult stem cells that can also be divided, basically, into two types: hematopoietic stem cells (HSCs) and mesenchymal stem cells (MSCs), originating hematopoietic and mesenchymal lineages, respectively [84, 85]. Both groups represent cellular sources that can be easily obtained and isolated from bone marrow aspirates, in addition to be an autologous source for therapies [22–24].

While HSCs originate in blood cells, MSCs can differentiate into various cell types of mesenchymal origin, including osteocytes, chondrocytes, adipocytes and myocytes [86]. In addition, the latter type of stem cells has an important supporting role (stroma) for HSCs in the bone marrow [86]. The mechanisms of action of HSCs and MSCs when they come into contact with injured tissue are not yet fully understood. It is currently suggested that these cell types can differentiate into glial and even neural lines [86]. Some studies mention that they can form glial and neuronal cells from various inducing mechanisms, such as chemical, genetic and physiological manipulations [86].

Other studies emphasize the trophic functions of these two cell types. It is known that HSCs can secrete neurotrophic growth factors such as angiopoietin-1, which has an angiogenic function [87]. There are also reports that MSCs may have an immunosuppressive function, which can reduce the acute inflammatory response, as well as reduce the reactivity of activated microglia/macrophages and astrocytes [22, 23, 88]. In addition, MSCs can promote axonal regeneration or positively influence functional plasticity through the modulation of an inflammatory medium that allows axonal growth [89]. They can synthesize some neurotrophic cytokines that stimulate neural growth, including BDNF (brain-derived neurotrophic factor), NGF (neural growth factor and VEGF (vascular endothelial growth factor).

Mesenchymal stem cells and bone marrow mononuclear cells promote improvement of functional deficits in animal models of stroke when administered intravenously, intra-arterially and intra-cerebrally [22, 23, 88, 90, 91], although most the injected cells to non-neural organs, mainly spleen [92–96].

Evidence from preclinical studies indicates that the main mechanism of cell therapy does not correspond to cell replacement directly, but to the trophic, anti-inflammatory and immunomodulatory effects that occur in the acute phase and that persist until the transplanted cells die [22, 23, 88, 90, 91].

The route of administration of these cells can be determined by choosing the time for transplantation, according to the therapeutic purpose. For example, intravascular transplants may require earlier delivery as the cells use acute inflammatory signals to reach the injured area [92–96]. On the other hand, intra-parenchymal injection could be beneficial in a later administration in order to favor the survival of these cells since the acute inflammatory environment causes damage to the transplanted cells [92–96].

6. Minocycline and neuroprotective actions

Minocycline is a second generation semi-synthetic tetracycline, commonly used as an antibiotic, but which has a considerable anti-inflammatory and neuroprotective effect in experimental models of stroke and trauma [46, 50, 51, 97, 98]. This has been first demonstrated by Yrjanheikki and colleagues using experimental models of both global [50] and focal [51] ischemia. Following MCAO in rats, minocycline treatment induced a 65% decrease in the cortical infarct area and a 45% reduction in the primary ischemic area [51]. The authors attributed these effects mainly to inhibition of microglial activation. From these initial studies, several other studies have shown the neuroprotective effects of minocycline after ischemia and several other diseases in the CNS [98–100].

The treatment of rodents submitted to acute SCI with minocycline reduced secondary oligodendrocyte degeneration, increased axonal regeneration and modulated cell death due to apoptosis [46]. Minocycline treatment increases endogenous neurogenesis in the adult brain after experimental stroke [101]. We have shown that minocycline protects striatal white matter following acute excitotoxic brain injury [102] and that modulation of microglial activation enhances the therapeutic actions of BMMCs into the acute phase of experimental stroke [22, 23].

Part of the success of minocycline may be associated with the chemical structure of this drug [103]. The molecular organization of minocycline allows it to be up to 5 times more lipophilic than the other tetracyclines [99, 103]. This facilitates the molecule to easy crosses the blood brain barrier (BBB) [99, 103]. In addition, minocycline is quickly and easily absorbed, well tolerated in high doses and has an average half-life superior to other drugs with similar biological action [56, 99, 103]. These characteristics make minocycline a therapeutic promise for several CNS diseases, including ischemic stroke {Yong, 2004 #917. These characteristics make minocycline a therapeutic promise for several CNS diseases, including ischemic stroke {Yong, 2004 #917},]104].

Although the mechanism of action of minocycline in ischemic stroke is not fully elucidated, the drug appears to exert influence on different points of the inflammatory response and apoptosis [56, 99, 104]. Minocycline blocks leukocyte activation and infiltration, attenuates the permeability of BBB, inhibits matrix metalloproteinase (MMPs), induced nitric oxide enzyme (iNOS), modulates inflammatory mediators, reduces microglial activation and proliferation [56, 99, 104, 105]. In addition, it has been reported that minocycline inhibits microglial activation by a specific action in a cytokine-like mediator called high-mobility group box-1 (HMGB-1) [106].

In the apoptotic cascade, minocycline can play a role on the extracellular availability of death ligands and/or in the presence of neurotrophic factors in the extracellular medium that activate survival receptors in the cell [107–109]. Intracellularly, the main target of minocycline is the mitochondria. In this organelle, the drug stabilizes the mitochondrial membrane and prevents the release of the enzyme cytochrome-c and downstream caspase-3 activation [107–110].

7. Modulation of microglia activation with minocycline to enhance neuroprotection after BMMC transplants

There is an issue on what is the best time window to transplant stem/progenitor cells after acute neural disorders as the intense inflammatory present in the pathological environment might impair survival of the transplanted cells [27–28]. It has been shown that an exacerbated immune/inflammatory response may impair survival of stem cells transplanted in both normal [27] and pathological tissue [28]. This has been observed in non-neural tissue, as in the case for transplants of exogenous stem cells for myocardial repair [111].

Recent studies using a neuronal relay approach for spinal cord injury (SCI) have considered possible detrimental effects of inflammatory response on fetal [112], embryonic [113] and even induced-pluripotent stem cells (IPSCs) [114]. In this experimental paradigm authors transplant the stem/progenitor cells only 10 days after experimental trauma to avoid the detrimental effects of inflammatory reaction on the transplanted neural progenitor cells [112–118].

It has been confirmed that uncontrolled activated microglia may be detrimental contributing to bystander neuronal damage after stroke [9, 10]. We raised the hypothesis that modulation of microglial activation in the ischemic environment would enhance the therapeutic effects of BMMCs transplanted into the acute phase of both cortical [22] and striatal [23] stroke. We then transplanted BMMCs into the acute phase of stroke in adult rats and concomitantly treated ischemic rats with minocycline during six days after BMMC transplants at 24 h from stroke induction [22–23]. We compared ischemic animals concomitantly treated with BMMCs and minocycline with animals treated with minocycline or BMMCs [22–23].

The results have shown that concomitant treatment of ischemic animals with BMMCs and minocycline afforded better neuroprotection and functional recovery (**Figures 2–3**) than single treatment with BMMCs or minocyline alone [22–23]. In this experimental paradigm, modulation of microglial activation with minocycline into the acute phase stroke improved the therapeutic actions of both BMMCs and minocycline indicating a therapeutic synergism. The results also suggest that exacerbated microglial activation may impair the therapeutic actions of stem cells transplanted into the acute phase of stroke [22–23].

We further confirmed the suitability of both BMMCs and minocycline as neuroprotective agents using an intracerebral route of transplantation in an experimental models of striatal stroke [24]. We have shown that the direct brain injection of BMMCs into the acute phase of striatal stroke induces better neuroprotection and functional recovery than the intravenous route, although this experimental approach is less invasive the surgical intra-striatal injection [22–24]. In the same study, we have obtained very important information on the peculiarities of minocycline and BMMCs as neuroprotective agents [24]. Both BMMCs and minocycline reduced the number of ED1+ cells, but BMMCs were more effective in reducing it. BMMCs also induced a more pronounced reduction in the number of apoptotic cells (active caspase+ cells) than minocycline. Both treatments were equally effect in reducing neuronal loss [24].

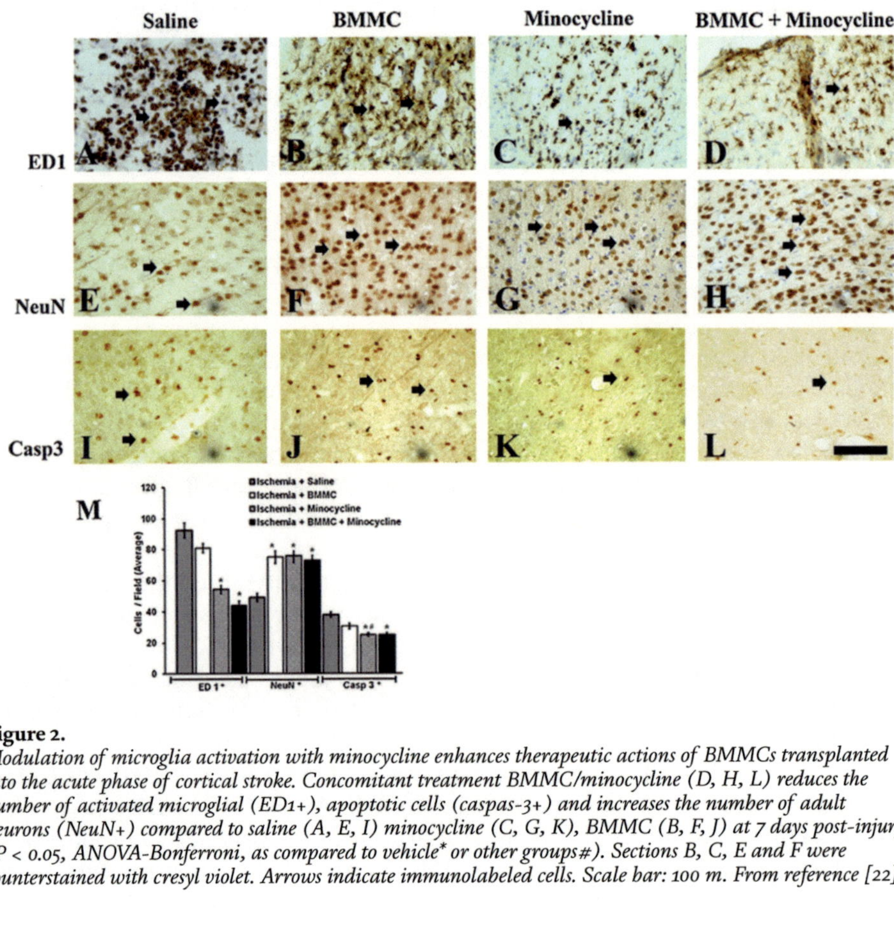

Figure 2.

Modulation of microglia activation with minocycline enhances therapeutic actions of BMMCs transplanted into the acute phase of cortical stroke. Concomitant treatment BMMC/minocycline (D, H, L) reduces the number of activated microglial (ED1+), apoptotic cells (caspas-3+) and increases the number of adult neurons (NeuN+) compared to saline (A, E, I) minocycline (C, G, K), BMMC (B, F, J) at 7 days post-injury. (P < 0.05, ANOVA-Bonferroni, as compared to vehicle or other groups#). Sections B, C, E and F were counterstained with cresyl violet. Arrows indicate immunolabeled cells. Scale bar: 100 m. From reference [22].*

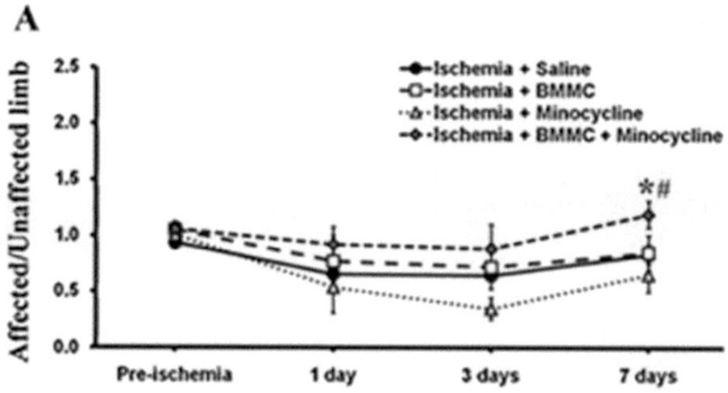

Figure 3.
Concomitant minocycline/BMMC treatment enhances the functional recovery induced by BMMCs into the acute phase of cortical stroke. BMMC-minocycline-treated animals showed better performance in the modified sticky-tape test than animals treated with BMMC or minocycline alone at 7 days post-injury ($P < 0.05$, ANOVA-Bonferroni, as compared to vehicle control animals). From reference [22].

The results suggest that modulation that minocycline and BMMCs are promising neuroprotective agents for experimental stroke and their concomitant use affords better neuroprotection and functional recovery than their single used [22–23]. In addition, intracerebral injections afford better therapeutic actions for BMMCs, although this experimental procedure is more invasive than the intravenous route [24]. The therapeutic synergism of the concomitant use of minocycline and BMMC is an important rationale to be explored in future investigations and a promising therapy for human stroke. It points out to the fact that a proper modulation of an exacerbated neuroinflammation in the ischemic environment is a suitable approach to enhance neuroprotection following transplants of stem cells into the acute phase of stroke and trauma.

8. Conclusion

In this chapter, we reviewed the pathophysiology of experimental stroke and the use of BMMCs as a promising approach to afford neuroprotection and functional recovery after transplants into the acute phase of brain ischemia. We have emphasized that transplanted progenitor/stem cells are affected by the pathological environment, including an exacerbated neuroinflammation. We have shown that a proper modulation of microglial activation of minocycline enhances both neuroprotection and functional recovery of BMMCs transplanted at 24 h after both cortical and striatal experimental stroke [22–23]. This approach can be used as an adjuvant therapy to enhance survival and efficacy of different kind of stem cells transplanted into the acute phase of stroke. In addition, this would reduce the time window of transplantation, which can be very important in the case of stroke, an acute neural disorder in which damage develops quickly.

Acknowledgements

This work was supported by the Brazilian National Council for Scientific and Technological Development (CNPq) and Fundação de Amparo A Pesquisa do Estado do Pará (FAPESPA). Authors thanks to Rosalia Mendez-Otero (UFRJ) and Carlos Alexandre Netto (UFSC) for technical support.

Conflict of interest

The authors declare no conflict of interest.

Author details

Edna Cristina S. Franco[1], Marcelo Marques Cardoso[2], Celice Cordeiro de Souza[3], Michelle Castro da Silva[4], Carolina Ramos dos Santos[2] and Walace Gomes-Leal[2]*

1 Pathology Unit, Evandro Chagas Institute, Belém, Pará, Brazil

2 Laboratory of Experimental Neuroprotection and Neuroregeneration, Institute of Collective Health, Federal University of Western Pará, Brazil

3 University Center of Pará (CESUPA), Belém, Para, Brazil

4 Laboratory of Experimental Neuroprotection and Neuroregeneration, Federal University of Pará, Para State University, Belem, Para, Brazil

*Address all correspondence to: wgomesleal@gmail.com

References

[1] Lo EH, Dalkara T, Moskowitz MA: Mechanisms, challenges and opportunities in stroke. Nat Rev Neurosci 2003; 4: 399-415. 10.1038/nrn1106:

[2] Chamorro A, Lo EH, Renu A, van Leyden K, Lyden PD: The future of neuroprotection in stroke. J Neurol Neurosurg Psychiatry 2020. 10.1136/jnnp-2020-324283:

[3] Kaiser EE, West FD: Large animal ischemic stroke models: replicating human stroke pathophysiology. Neural Regen Res 2020; 15: 1377-87. 10.4103/1673-5374.274324:

[4] Powers WJ: Acute Ischemic Stroke. N Engl J Med 2020; 383: 252-60. 10.1056/NEJMcp1917030:

[5] Moskowitz MA, Lo EH, Iadecola C: The science of stroke: mechanisms in search of treatments. Neuron 2010; 67: 181-98. 10.1016/j.neuron.2010.07.002:

[6] Wafa HA, Wolfe CDA, Emmett E, Roth GA, Johnson CO, Wang Y: Burden of Stroke in Europe: Thirty-Year Projections of Incidence, Prevalence, Deaths, and Disability-Adjusted Life Years. Stroke 2020; 51: 2418-27. 10.1161/STROKEAHA.120.029606:

[7] Virani SS, Alonso A, Benjamin EJ, Bittencourt MS, Callaway CW, Carson AP, Chamberlain AM, Chang AR, Cheng S, Delling FN, Djousse L, Elkind MSV, Ferguson JF, Fornage M, Khan SS, Kissela BM, Knutson KL, Kwan TW, Lackland DT, Lewis TT, Lichtman JH, Longenecker CT, Loop MS, Lutsey PL, Martin SS, Matsushita K, Moran AE, Mussolino ME, Perak AM, Rosamond WD, Roth GA, Sampson UKA, Satou GM, Schroeder EB, Shah SH, Shay CM, Spartano NL, Stokes A, Tirschwell DL, VanWagner LB, Tsao CW, American Heart Association Council on E, Prevention Statistics C, Stroke Statistics S: Heart Disease and Stroke Statistics-2020 Update: A Report From the American Heart Association. Circulation 2020; 141: e139-e596. 10.1161/CIR.0000000000000757:

[8] Jiao Y, Liu YW, Chen WG, Liu J: Neuroregeneration and functional recovery after stroke: advancing neural stem cell therapy toward clinical application. Neural Regen Res 2021; 16: 80-92. 10.4103/1673-5374.286955:

[9] Gomes-Leal W: Why microglia kill neurons after neural disorders? The friendly fire hypothesis. Neural Regen Res 2019; 14: 1499-502. 10.4103/1673-5374.255359:

[10] Gomes-Leal W: Microglial physiopathology: how to explain the dual role of microglia after acute neural disorders? Brain Behav 2012; 2: 345-56. 10.1002/brb3.51:

[11] Carpenter CR, Keim SM, Milne WK, Meurer WJ, Barsan WG, Best Evidence in Emergency Medicine Investigator G: Thrombolytic therapy for acute ischemic stroke beyond three hours. J Emerg Med 2011; 40: 82-92. 10.1016/j.jemermed.2010.05.009:

[12] Kablau M, Alonso A, Hennerici MG, Fatar M: Treatment with tPA predicts better outcome even if MCA occlusion persists. Int J Stroke 2013; 8: 496-502. 10.1111/j.1747-4949.2011.00750.x:

[13] Kablau M, Kreisel SH, Sauer T, Binder J, Szabo K, Hennerici MG, Kern R: Predictors and early outcome of hemorrhagic transformation after acute ischemic stroke. Cerebrovasc Dis 2011; 32: 334-41. 10.1159/000331702:

[14] Ginsberg MD: Current status of neuroprotection for cerebral ischemia: synoptic overview. Stroke 2009; 40: S111-4. 10.1161/STROKEAHA.108.528877:

[15] Ginsberg MD: Expanding the concept of neuroprotection for acute

ischemic stroke: The pivotal roles of reperfusion and the collateral circulation. Prog Neurobiol 2016; 145-146: 46-77. 10.1016/j.pneurobio.2016.09.002:

[16] Lindvall O, Kokaia Z: Stem cell therapy for human brain disorders. Kidney Int 2005; 68: 1937-9. 10.1111/j.1523-1755.2005.00623.x:

[17] Kokaia Z, Darsalia V: Human Neural Stem Cells for Ischemic Stroke Treatment. Results Probl Cell Differ 2018; 66: 249-63. 10.1007/978-3-319-93485-3_11:

[18] Kokaia Z, Llorente IL, Carmichael ST: Customized Brain Cells for Stroke Patients Using Pluripotent Stem Cells. Stroke 2018; 49: 1091-8. 10.1161/STROKEAHA.117.018291:

[19] Gronning Hansen M, Laterza C, Palma-Tortosa S, Kvist G, Monni E, Tsupykov O, Tornero D, Uoshima N, Soriano J, Bengzon J, Martino G, Skibo G, Lindvall O, Kokaia Z: Grafted human pluripotent stem cell-derived cortical neurons integrate into adult human cortical neural circuitry. Stem Cells Transl Med 2020. 10.1002/sctm.20-0134:

[20] Palma-Tortosa S, Tornero D, Gronning Hansen M, Monni E, Hajy M, Kartsivadze S, Aktay S, Tsupykov O, Parmar M, Deisseroth K, Skibo G, Lindvall O, Kokaia Z: Activity in grafted human iPS cell-derived cortical neurons integrated in stroke-injured rat brain regulates motor behavior. Proc Natl Acad Sci U S A 2020; 117: 9094-100. 10.1073/pnas.2000690117:

[21] Lindvall O, Kokaia Z: Stem cells for the treatment of neurological disorders. Nature 2006; 441: 1094-6. 10.1038/nature04960:

[22] Franco EC, Cardoso MM, Gouveia A, Pereira A, Gomes-Leal W: Modulation of microglial activation enhances neuroprotection and functional recovery derived from bone marrow mononuclear cell transplantation after cortical ischemia. Neurosci Res 2012; 73: 122-32. 10.1016/j.neures.2012.03.006:

[23] Cardoso MM, Franco EC, de Souza CC, da Silva MC, Gouveia A, Gomes-Leal W: Minocycline treatment and bone marrow mononuclear cell transplantation after endothelin-1 induced striatal ischemia. Inflammation 2013; 36: 197-205. 10.1007/s10753-012-9535-5:

[24] Souza CC, da Silva MC, Lopes RT, Cardoso MM, de Souza LL, Santos AG, Dos Santos IR, Franco ECS, Gomes-Leal W: Comparative Therapeutic Effects of Minocycline Treatment and Bone Marrow Mononuclear Cell Transplantation following Striatal Stroke. Oxid Med Cell Longev 2017; 2017: 1976191. 10.1155/2017/1976191:

[25] Chumnanvej S, Chumnanvej S: Autologous bone-marrow mononuclear stem cell therapy in patients with stroke: a meta-analysis of comparative studies. Biomed Eng Online 2020; 19: 74. 10.1186/s12938-020-00819-7:

[26] Kawabori M, Shichinohe H, Kuroda S, Houkin K: Clinical Trials of Stem Cell Therapy for Cerebral Ischemic Stroke. Int J Mol Sci 2020; 21. 10.3390/ijms21197380:

[27] Coyne TM, Marcus AJ, Woodbury D, Black IB: Marrow stromal cells transplanted to the adult brain are rejected by an inflammatory response and transfer donor labels to host neurons and glia. Stem Cells 2006; 24: 2483-92. 10.1634/stemcells.2006-0174:

[28] Molcanyi M, Riess P, Bentz K, Maegele M, Hescheler J, Schafke B, Trapp T, Neugebauer E, Klug N, Schafer U: Trauma-associated inflammatory response impairs embryonic stem cell survival and integration after implantation into injured rat brain. J

Neurotrauma 2007; 24: 625-37. 10.1089/neu.2006.0180:

[29] Dos Santos IRC, Dias MNC, Gomes-Leal W: Microglial activation and adult neurogenesis after brain stroke. Neural Regen Res 2021; 16: 456-9. 10.4103/1673-5374.291383:

[30] Puig B, Brenna S, Magnus T: Molecular Communication of a Dying Neuron in Stroke. Int J Mol Sci 2018; 19. 10.3390/ijms19092834:

[31] Fang Y, Gao S, Wang X, Cao Y, Lu J, Chen S, Lenahan C, Zhang JH, Shao A, Zhang J: Programmed Cell Deaths and Potential Crosstalk With Blood-Brain Barrier Dysfunction After Hemorrhagic Stroke. Front Cell Neurosci 2020; 14: 68. 10.3389/fncel.2020.00068:

[32] Datta A, Sarmah D, Mounica L, Kaur H, Kesharwani R, Verma G, Veeresh P, Kotian V, Kalia K, Borah A, Wang X, Dave KR, Yavagal DR, Bhattacharya P: Cell Death Pathways in Ischemic Stroke and Targeted Pharmacotherapy. Transl Stroke Res 2020; 11: 1185-202. 10.1007/s12975-020-00806-z:

[33] Sekerdag E, Solaroglu I, Gursoy-Ozdemir Y: Cell Death Mechanisms in Stroke and Novel Molecular and Cellular Treatment Options. Curr Neuropharmacol 2018; 16: 1396-415. 10.2174/1570159X16666180302115544:

[34] Yuan J: Neuroprotective strategies targeting apoptotic and necrotic cell death for stroke. Apoptosis 2009; 14: 469-77. 10.1007/s10495-008-0304-8:

[35] Liu C, Zhang K, Shen H, Yao X, Sun Q, Chen G: Necroptosis: A novel manner of cell death, associated with stroke (Review). Int J Mol Med 2018; 41: 624-30. 10.3892/ijmm.2017.3279:

[36] Bergsbaken T, Fink SL, Cookson BT: Pyroptosis: host cell death and inflammation. Nat Rev Microbiol 2009; 7: 99-109. 10.1038/nrmicro2070:

[37] Iadecola C, Buckwalter MS, Anrather J: Immune responses to stroke: mechanisms, modulation, and therapeutic potential. J Clin Invest 2020; 130: 2777-88. 10.1172/JCI135530:

[38] Dos Santos CD, Picanco-Diniz CW, Gomes-Leal W: Differential patterns of inflammatory response, axonal damage and myelin impairment following excitotoxic or ischemic damage to the trigeminal spinal nucleus of adult rats. Brain Res 2007; 1172: 130-44. 10.1016/j.brainres.2007.07.037:

[39] Gomes-Leal W, Corkill DJ, Freire MA, Picanco-Diniz CW, Perry VH: Astrocytosis, microglia activation, oligodendrocyte degeneration, and pyknosis following acute spinal cord injury. Exp Neurol 2004; 190: 456-67. 10.1016/j.expneurol.2004.06.028:

[40] Gomes-Leal W, Corkill DJ, Picanco-Diniz CW: Systematic analysis of axonal damage and inflammatory response in different white matter tracts of acutely injured rat spinal cord. Brain Res 2005; 1066: 57-70. 10.1016/j.brainres.2005.10.069:

[41] Lima RR, Santana LN, Fernandes RM, NascimentoEM, OliveiraAC, FernandesLM, Dos Santos EM, Tavares PA, Dos Santos IR, Gimaraes-Santos A, Gomes-Leal W: Neurodegeneration and Glial Response after Acute Striatal Stroke: Histological Basis for Neuroprotective Studies. Oxid Med Cell Longev 2016; 2016: 3173564. 10.1155/2016/3173564:

[42] Souza-Rodrigues RD, Costa AM, Lima RR, Dos Santos CD, Picanco-Diniz CW, Gomes-Leal W: Inflammatory response and white matter damage after microinjections of endothelin-1 into the rat striatum. Brain Res 2008; 1200: 78-88. 10.1016/j.brainres.2007.11.025:

[43] Schnell L, Fearn S, Klassen H, Schwab ME, Perry VH: Acute inflammatory responses to

mechanical lesions in the CNS: differences between brain and spinal cord. Eur J Neurosci 1999; 11: 3648-58. 10.1046/j.1460-9568.1999.00792.x:

[44] Tanaka R, Komine-Kobayashi M, Mochizuki H, Yamada M, Furuya T, Migita M, Shimada T, Mizuno Y, Urabe T: Migration of enhanced green fluorescent protein expressing bone marrow-derived microglia/macrophage into the mouse brain following permanent focal ischemia. Neuroscience 2003; 117: 531-9. 10.1016/s0306-4522(02)00954-5:

[45] Kronenberg G, Uhlemann R, Richter N, Klempin F, Wegner S, Staerck L, Wolf S, Uckert W, Kettenmann H, Endres M, Gertz K: Distinguishing features of microglia- and monocyte-derived macrophages after stroke. Acta Neuropathol 2018; 135: 551-68. 10.1007/s00401-017-1795-6:

[46] Stirling DP, Khodarahmi K, Liu J, McPhail LT, McBride CB, Steeves JD, Ramer MS, Tetzlaff W: Minocycline treatment reduces delayed oligodendrocyte death, attenuates axonal dieback, and improves functional outcome after spinal cord injury. J Neurosci 2004; 24: 2182-90. 10.1523/JNEUROSCI.5275-03.2004:

[47] Beattie MS: Inflammation and apoptosis: linked therapeutic targets in spinal cord injury. Trends Mol Med 2004; 10: 580-3. 10.1016/j.molmed.2004.10.006:

[48] Casha S, Zygun D, McGowan MD, Bains I, Yong VW, Hurlbert RJ: Results of a phase II placebo-controlled randomized trial of minocycline in acute spinal cord injury. Brain 2012; 135: 1224-36. 10.1093/brain/aws072:

[49] Koistinaho J, Yrjanheikki J, Kauppinen T, Koistinaho M: Tetracycline derivatives as anti-inflammatory agents and potential agents in stroke treatment. Ernst Schering Res Found Workshop 2004; 101-15. 10.1007/978-3-662-05426-0_6:

[50] Yrjanheikki J, Keinanen R, Pellikka M, Hokfelt T, Koistinaho J: Tetracyclines inhibit microglial activation and are neuroprotective in global brain ischemia. Proc Natl Acad Sci U S A 1998; 95: 15769-74. 10.1073/pnas.95.26.15769:

[51] Yrjanheikki J, Tikka T, Keinanen R, Goldsteins G, Chan PH, Koistinaho J: A tetracycline derivative, minocycline, reduces inflammation and protects against focal cerebral ischemia with a wide therapeutic window. Proc Natl Acad Sci U S A 1999; 96: 13496-500. 10.1073/pnas.96.23.13496:

[52] Popovich PG, Guan Z, Wei P, Huitinga I, van Rooijen N, Stokes BT: Depletion of hematogenous macrophages promotes partial hindlimb recovery and neuroanatomical repair after experimental spinal cord injury. Exp Neurol 1999; 158: 351-65. 10.1006/exnr.1999.7118:

[53] Fagan SC: Drug repurposing for drug development in stroke. Pharmacotherapy 2010; 30: 51S–4S. 10.1592/phco.30.pt2.51S:

[54] Fouda AY, Newsome AS, Spellicy S, Waller JL, Zhi W, Hess DC, Ergul A, Edwards DJ, Fagan SC, Switzer JA: Minocycline in Acute Cerebral Hemorrhage: An Early Phase Randomized Trial. Stroke 2017; 48: 2885-7. 10.1161/STROKEAHA.117.018658:

[55] Lampl Y, Boaz M, Gilad R, Lorberboym M, Dabby R, Rapoport A, Anca-Hershkowitz M, Sadeh M: Minocycline treatment in acute stroke: an open-label, evaluator-blinded study. Neurology 2007; 69: 1404-10. 10.1212/01.wnl.0000277487.04281.db:

[56] Malhotra K, Chang JJ, Khunger A, Blacker D, Switzer JA, Goyal N, Hernandez AV, Pasupuleti V, A lexandrov AV, Tsivgoulis G: Minocycline

for acute stroke treatment: a systematic review and meta-analysis of randomized clinical trials. J Neurol 2018; 265: 1871-9. 10.1007/s00415-018-8935-3:

[57] Radlinska BA, Ghinani SA, Lyon P, Jolly D, Soucy JP, Minuk J, Schirrmacher R, Thiel A: Multimodal microglia imaging of fiber tracts in acute subcortical stroke. Ann Neurol 2009; 66: 825-32. 10.1002/ana.21796:

[58] Thiel A, Heiss WD: Imaging of microglia activation in stroke. Stroke 2011; 42: 507-12. 10.1161/STROKEAHA.110.598821:

[59] Thiel A, Radlinska BA, Paquette C, Sidel M, Soucy JP, Schirrmacher R, Minuk J: The temporal dynamics of poststroke neuroinflammation: a longitudinal diffusion tensor imaging-guided PET study with 11C-PK11195 in acute subcortical stroke. J Nucl Med 2010; 51: 1404-12. 10.2967/jnumed.110.076612:

[60] Shechter R, London A, Varol C, Raposo C, Cusimano M, Yovel G, Rolls A, Mack M, Pluchino S, Martino G, Jung S, Schwartz M: Infiltrating blood-derived macrophages are vital cells playing an anti-inflammatory role in recovery from spinal cord injury in mice. PLoS Med 2009; 6: e1000113. 10.1371/journal.pmed.1000113:

[61] Shechter R, Schwartz M: Harnessing monocyte-derived macrophages to control central nervous system pathologies: no longer 'if' but 'how'. J Pathol 2013; 229: 332-46. 10.1002/path.4106:

[62] Kigerl KA, Gensel JC, Ankeny DP, Alexander JK, Donnelly DJ, Popovich PG: Identification of two distinct macrophage subsets with divergent effects causing either neurotoxicity or regeneration in the injured mouse spinal cord. J Neurosci 2009; 29: 13435-44. 10.1523/JNEUROSCI.3257-09.2009:

[63] Rapalino O, Lazarov-Spiegler O, Agranov E, Velan GJ, Yoles E, Fraidakis M, Solomon A, Gepstein R, Katz A, Belkin M, Hadani M, Schwartz M: Implantation of stimulated homologous macrophages results in partial recovery of paraplegic rats. Nat Med 1998; 4: 814-21. 10.1038/nm0798-814:

[64] Neumann H, Kotter MR, Franklin RJ: Debris clearance by microglia: an essential link between degeneration and regeneration. Brain 2009; 132: 288-95. 10.1093/brain/awn109:

[65] Neumann J, Gunzer M, Gutzeit HO, Ullrich O, Reymann KG, Dinkel K: Microglia provide neuroprotection after ischemia. FASEB J 2006; 20: 714-6. 10.1096/fj.05-4882fje:

[66] Neumann J, Sauerzweig S, Ronicke R, Gunzer F, Dinkel K, Ullrich O, Gunzer M, Reymann KG: Microglia cells protect neurons by direct engulfment of invading neutrophil granulocytes: a new mechanism of CNS immune privilege. J Neurosci 2008; 28: 5965-75. 10.1523/JNEUROSCI.0060-08.2008:

[67] Lalancette-Hebert M, Gowing G, Simard A, Weng YC, Kriz J: Selective ablation of proliferating microglial cells exacerbates ischemic injury in the brain. J Neurosci 2007; 27: 2596-605. 10.1523/JNEUROSCI.5360-06.2007:

[68] Rayasam A, Hsu M, Kijak JA, Kissel L, Hernandez G, Sandor M, Fabry Z: Immune responses in stroke: how the immune system contributes to damage and healing after stroke and how this knowledge could be translated to better cures? Immunology 2018; 154: 363-76. 10.1111/imm.12918:

[69] Xu Y, Jin MZ, Yang ZY, Jin WL: Microglia in neurodegenerative diseases. Neural Regen Res 2021; 16: 270-80. 10.4103/1673-5374.290881:

[70] Zhang W, Tian T, Gong SX, Huang WQ, Zhou QY, Wang AP, Tian Y: Microglia-associated neuroinflammation is a potential therapeutic target for

ischemic stroke. Neural Regen Res 2021; 16: 6-11. 10.4103/1673-5374.286954:

[71] Lambertsen KL, Finsen B, Clausen BH: Post-stroke inflammation-target or tool for therapy? Acta Neuropathol 2019; 137: 693-714. 10.1007/s00401-018-1930-z:

[72] Ginhoux F, Garel S: The mysterious origins of microglia. Nat Neurosci 2018; 21: 897-9. 10.1038/s41593-018-0176-3:

[73] Ginhoux F, Greter M, Leboeuf M, Nandi S, See P, Gokhan S, Mehler MF, Conway SJ, Ng LG, Stanley ER, Samokhvalov IM, Merad M: Fate mapping analysis reveals that adult microglia derive from primitive macrophages. Science 2010; 330: 841-5. 10.1126/science.1194637:

[74] Nimmerjahn A: Monitoring neuronal health. Science 2020; 367: 510-1. 10.1126/science.aba4472:

[75] Nimmerjahn A, Kirchhoff F, Helmchen F: Resting microglial cells are highly dynamic surveillants of brain parenchyma in vivo. Science 2005; 308: 1314-8. 10.1126/science.1110647:

[76] Davalos D, Grutzendler J, Yang G, Kim JV, Zuo Y, Jung S, Littman DR, Dustin ML, Gan WB: ATP mediates rapid microglial response to local brain injury in vivo. Nat Neurosci 2005; 8: 752-8. 10.1038/nn1472:

[77] Vainchtein ID, Chin G, Cho FS, Kelley KW, Miller JG, Chien EC, Liddelow SA, Nguyen PT, Nakao-Inoue H, Dorman LC, Akil O, Joshita S, Barres BA, Paz JT, Molofsky AB, Molofsky AV: Astrocyte-derived interleukin-33 promotes microglial synapse engulfment and neural circuit development. Science 2018; 359: 1269-73. 10.1126/science.aal3589:

[78] Barichello T, Generoso JS, Simoes LR, Goularte JA, Petronilho F, Saigal P, Badawy M, Quevedo J: Role of Microglial Activation in the Pathophysiology of Bacterial Meningitis. Mol Neurobiol 2016; 53: 1770-81. 10.1007/s12035-015-9107-4:

[79] Shastri A, Bonifati DM, Kishore U: Innate immunity and neuro-inflammation. Mediators Inflamm 2013; 2013: 342931. 10.1155/2013/342931:

[80] Spanos JP, Hsu NJ, Jacobs M: Microglia are crucial regulators of neuro-immunity during central nervous system tuberculosis. Front Cell Neurosci 2015; 9: 182. 10.3389/fncel.2015.00182:

[81] Sacchetti ML: Is it time to definitely abandon neuroprotection in acute ischemic stroke? Stroke 2008; 39: 1659-60. 10.1161/STROKEAHA.107.505024:

[82] Jaillard A, Hommel M, Moisan A, Zeffiro TA, Favre-Wiki IM, Barbieux-Guillot M, Vadot W, Marcel S, Lamalle L, Grand S, Detante O: Autologous Mesenchymal Stem Cells Improve Motor Recovery in Subacute Ischemic Stroke: a Randomized Clinical Trial. Transl Stroke Res 2020; 11: 910-23. 10.1007/s12975-020-00787-z:

[83] Mendez-Otero R, de Freitas GR, Andre C, de Mendonca ML, Friedrich M, Oliveira-Filho J: Potential roles of bone marrow stem cells in stroke therapy. Regen Med 2007; 2: 417-23. 10.2217/17460751.2.4.417:

[84] Lucas D: Structural organization of the bone marrow and its role in hematopoiesis. Curr Opin Hematol 2020. 10.1097/MOH.0000000000000621:

[85] Hurwitz SN, Jung SK, Kurre P: Hematopoietic stem and progenitor cell signaling in the niche. Leukemia 2020. 10.1038/s41375-020-01062-8:

[86] Krabbe C, Zimmer J, Meyer M: Neural transdifferentiation of mesenchymal stem cells--a critical review. APMIS 2005; 113: 831-44. 10.1111/j.1600-0463.2005.apm_3061.x:

[87] Takakura N, Watanabe T, Suenobu S, Yamada Y, Noda T, Ito Y, Satake M, Suda T: A role for hematopoietic stem cells in promoting angiogenesis. Cell 2000; 102: 199-209. 10.1016/s0092-8674(00)00025-8:

[88] Giraldi-Guimardes A, Rezende-Lima M, Bruno FP, Mendez-Otero R: Treatment with bone marrow mononuclear cells induces functional recovery and decreases neurodegeneration after sensorimotor cortical ischemia in rats. Brain Res 2009; 1266: 108-20. 10.1016/j.brainres.2009.01.062:

[89] Zhukareva V, Obrocka M, Houle JD, Fischer I, Neuhuber B: Secretion profile of human bone marrow stromal cells: donor variability and response to inflammatory stimuli. Cytokine 2010; 50: 317-21. 10.1016/j.cyto.2010.01.004:

[90] de Vasconcelos Dos Santos A, da Costa Reis J, Diaz Paredes B, Moraes L, Jasmin, Giraldi-Guimaraes A, Mendez-Otero R: Therapeutic window for treatment of cortical ischemia with bone marrow-derived cells in rats. Brain Res 2010; 1306: 149-58. 10.1016/j.brainres.2009.09.094:

[91] Vasconcelos-dos-Santos A, Rosado-de-Castro PH, Lopes de Souza SA, da Costa Silva J, Ramos AB, Rodriguez de Freitas G, Barbosa da Fonseca LM, Gutfilen B, Mendez-Otero R: Intravenous and intra-arterial administration of bone marrow mononuclear cells after focal cerebral ischemia: Is there a difference in biodistribution and efficacy? Stem Cell Res 2012; 9: 1-8. 10.1016/j.scr.2012.02.002:

[92] Keimpema E, Fokkens MR, Nagy Z, Agoston V, Luiten PG, Nyakas C, Boddeke HW, Copray JC: Early transient presence of implanted bone marrow stem cells reduces lesion size after cerebral ischaemia in adult rats. Neuropathol Appl Neurobiol 2009; 35: 89-102. 10.1111/j.1365-2990.2008.00961.x:

[93] Barbosa da Fonseca LM, Gutfilen B, Rosado de Castro PH, Battistella V, Goldenberg RC, Kasai-Brunswick T, Chagas CL, Wajnberg E, Maiolino A, Salles Xavier S, Andre C, Mendez-Otero R, de Freitas GR: Migration and homing of bone-marrow mononuclear cells in chronic ischemic stroke after intra-arterial injection. Exp Neurol 2010; 221: 122-8. 10.1016/j.expneurol.2009.10.010:

[94] Correa PL, Mesquita CT, Felix RM, Azevedo JC, Barbirato GB, Falcao CH, Gonzalez C, Mendonca ML, Manfrim A, de Freitas G, Oliveira CC, Silva D, Avila D, Borojevic R, Alves S, Oliveira AC, Jr., Dohmann HF: Assessment of intra-arterial injected autologous bone marrow mononuclear cell distribution by radioactive labeling in acute ischemic stroke. Clin Nucl Med 2007; 32: 839-41. 10.1097/RLU.0b013e318156b980:

[95] Rosado-de-Castro PH, Schmidt Fda R, Battistella V, Lopes de Souza SA, Gutfilen B, Goldenberg RC, Kasai-Brunswick TH, Vairo L, Silva RM, Wajnberg E, Alvarenga Americano do Brasil PE, Gasparetto EL, Maiolino A, Alves-Leon SV, Andre C, Mendez-Otero R, Rodriguez de Freitas G, Barbosa da Fonseca LM: Biodistribution of bone marrow mononuclear cells after intra-arterial or intravenous transplantation in subacute stroke patients. Regen Med 2013; 8: 145-55. 10.2217/rme.13.2:

[96] Yang B, Strong R, Sharma S, Brenneman M, Mallikarjunarao K, Xi X, Grotta JC, Aronowski J, Savitz SI: Therapeutic time window and dose response of autologous bone marrow mononuclear cells for ischemic stroke. J Neurosci Res 2011; 89: 833-9. 10.1002/jnr.22614:

[97] Dai S, Hua Y, Keep RF, Novakovic N, Fei Z, Xi G: Minocycline attenuates

brain injury and iron overload after intracerebral hemorrhage in aged female rats. Neurobiol Dis 2019; 126: 76-84. 10.1016/j.nbd.2018.06.001:

[98] Naderi Y, Panahi Y, Barreto GE, Sahebkar A: Neuroprotective effects of minocycline on focal cerebral ischemia injury: a systematic review. Neural Regen Res 2020; 15: 773-82. 10.4103/1673-5374.268898:

[99] Yong VW, Wells J, Giuliani F, Casha S, Power C, Metz LM: The promise of minocycline in neurology. Lancet Neurol 2004; 3: 744-51. 10.1016/S1474-4422(04)00937-8:

[100] Liao TV, Forehand CC, Hess DC, Fagan SC: Minocycline repurposing in critical illness: focus on stroke. Curr Top Med Chem 2013; 13: 2283-90. 10.2174/15680266113136660160:

[101] Liu Z, Fan Y, Won SJ, Neumann M, Hu D, Zhou L, Weinstein PR, Liu J: Chronic treatment with minocycline preserves adult new neurons and reduces functional impairment after focal cerebral ischemia. Stroke 2007; 38: 146-52. 10.1161/01.STR.0000251791.64910.cd:

[102] Guimaraes JS, Freire MA, Lima RR, Picanco-Diniz CW, Pereira A, Gomes-Leal W: Minocycline treatment reduces white matter damage after excitotoxic striatal injury. Brain Res 2010; 1329: 182-93. 10.1016/j.brainres.2010.03.007:

[103] Bahrami F, Morris DL, Pourgholami MH: Tetracyclines: drugs with huge therapeutic potential. Mini Rev Med Chem 2012; 12: 44-52. 10.2174/138955712798868977:

[104] Jiang Z, Alamuri TT, Muir ER, Choi DW, Duong TQ: Longitudinal multiparametric MRI study of hydrogen-enriched water with minocycline combination therapy in experimental ischemic stroke in rats. Brain Res 2020; 1748: 147122. 10.1016/j.brainres.2020.147122:

[105] Stirling DP, Koochesfahani KM, Steeves JD, Tetzlaff W: Minocycline as a neuroprotective agent. Neuroscientist 2005; 11: 308-22. 10.1177/1073858405275175:

[106] Hayakawa K, Mishima K, Nozako M, Hazekawa M, Mishima S, Fujioka M, Orito K, Egashira N, Iwasaki K, Fujiwara M: Delayed treatment with minocycline ameliorates neurologic impairment through activated microglia expressing a high-mobility group box1-inhibiting mechanism. Stroke 2008; 39: 951-8. 10.1161/STROKEAHA.107.495820:

[107] Zhu S, Stavrovskaya IG, Drozda M, Kim BY, Ona V, Li M, Sarang S, Liu AS, Hartley DM, Wu DC, Gullans S, Ferrante RJ, Przedborski S, Kristal BS, Friedlander RM: Minocycline inhibits cytochrome c release and delays progression of amyotrophic lateral sclerosis in mice. Nature 2002; 417: 74-8. 10.1038/417074a:

[108] Lu Y, Huang Z, Hua Y, Xiao G: Minocycline Promotes BDNF Expression of N2a Cells via Inhibition of miR-155-Mediated Repression After Oxygen-Glucose Deprivation and Reoxygenation. Cell Mol Neurobiol 2018; 38: 1305-13. 10.1007/s10571-018-0599-0:

[109] Domercq M, Matute C: Neuroprotection by tetracyclines. Trends Pharmacol Sci 2004; 25: 609-12. 10.1016/j.tips.2004.10.001:

[110] Jordan J, de Groot PW, Galindo MF: Mitochondria: the headquarters in ischemia-induced neuronal death. Cent Nerv Syst Agents Med Chem 2011; 11: 98-106. 10.2174/187152411796011358:

[111] Buja LM, Vela D: Immunologic and inflammatory reactions to exogenous stem cells implications for experimental studies and clinical trials for myocardial repair. J Am Coll Cardiol 2010; 56: 1693-700. 10.1016/j.jacc.2010.06.041:

[112] Lu P, Wang Y, Graham L, McHale K, Gao M, Wu D, Brock J, Blesch A, Rosenzweig ES, Havton LA, Zheng B, Conner JM, Marsala M, Tuszynski MH: Long-distance growth and connectivity of neural stem cells after severe spinal cord injury. Cell 2012; 150: 1264-73. 10.1016/j.cell.2012.08.020:

[113] Lu P, Kadoya K, Tuszynski MH: Axonal growth and connectivity from neural stem cell grafts in models of spinal cord injury. Curr Opin Neurobiol 2014; 27: 103-9. 10.1016/j.conb.2014.03.010:

[114] Rosenzweig ES, Brock JH, Lu P, Kumamaru H, Salegio EA, Kadoya K, Weber JL, Liang JJ, Moseanko R, Hawbecker S, Huie JR, Havton LA, Nout-Lomas YS, Ferguson AR, Beattie MS, Bresnahan JC, Tuszynski MH: Restorative effects of human neural stem cell grafts on the primate spinal cord. Nat Med 2018; 24: 484-90. 10.1038/nm.4502:

[115] Kadoya K, Lu P, Nguyen K, Lee-Kubli C, Kumamaru H, Yao L, Knackert J, Poplawski G, Dulin JN, Strobl H, Takashima Y, Biane J, Conner J, Zhang SC, Tuszynski MH: Spinal cord reconstitution with homologous neural grafts enables robust corticospinal regeneration. Nat Med 2016; 22: 479-87. 10.1038/nm.4066:

[116] Kumamaru H, Lu P, Rosenzweig ES, Kadoya K, Tuszynski MH: Regenerating Corticospinal Axons Innervate Phenotypically Appropriate Neurons within Neural Stem Cell Grafts. Cell Rep 2019; 26: 2329-39 e4. 10.1016/j.celrep.2019.01.099:

[117] Kumamaru H, Lu P, Rosenzweig ES, Tuszynski MH: Activation of Intrinsic Growth State Enhances Host Axonal Regeneration into Neural Progenitor Cell Grafts. Stem Cell Reports 2018; 11: 861-8. 10.1016/j.stemcr.2018.08.009:

[118] Lu P, Gomes-Leal W, Anil S, Dobkins G, Huie JR, Ferguson AR, Graham L, Tuszynski M: Origins of Neural Progenitor Cell-Derived Axons Projecting Caudally after Spinal Cord Injury. Stem Cell Reports 2019; 13: 105-14. 10.1016/j.stemcr.2019.05.011: